Moral Injury among Returning Veterans

Moral Injury among Returning Veterans

From Thank You for Your Service to a Liberative Solidarity

Joshua Morris

LEXINGTON BOOKS
Lanham • Boulder • New York • London

Published by Lexington Books
An imprint of The Rowman & Littlefield Publishing Group, Inc.
4501 Forbes Boulevard, Suite 200, Lanham, Maryland 20706
www.rowman.com

86-90 Paul Street, London EC2A 4NE

Copyright © 2021 by The Rowman & Littlefield Publishing Group, Inc.

All rights reserved. No part of this book may be reproduced in any form or by any electronic or mechanical means, including information storage and retrieval systems, without written permission from the publisher, except by a reviewer who may quote passages in a review.

British Library Cataloguing in Publication Information available

Library of Congress Cataloging-in-Publication Data

Names: Morris, Joshua T., 1982– author.
Title: Moral injury among returning veterans : from thank you for your service to a liberative solidarity / Joshua Morris.
Description: Lanham : Lexington Books, [2021] | Includes bibliographical references and index. | Summary: "This book expands the conversation on moral injury to include a more formal role for society in it. The author utilizes an interdisciplinary practical theology combining liberation theologies and cultural studies to interrogate how dominate ideologies can complicate moral injury reintegration among veterans"— Provided by publisher.
Identifiers: LCCN 2021032399 (print) | LCCN 2021032400 (ebook) | ISBN 9781793642646 (hardback) | ISBN 9781793642660 (epub) | ISBN 9781793642660 (pbk)
Subjects: LCSH: War—Moral and ethical aspects. | War—Psychological aspects. | Combat—Pyschological aspects. | Veterans—Mental health—United States. | Veteran reintegration—United States. | Post-traumatic stress disorder. | Theology, Practical.
Classification: LCC U22 .M67 2021 (print) | LCC U22 (ebook) | DDC 174/.9355—dc23
LC record available at https://lccn.loc.gov/2021032399
LC ebook record available at https://lccn.loc.gov/2021032400

ISBN 978-1-7936-4266-0 (paper)

Contents

Acknowledgments	vii
Chapter 1: Are We Still Over *There*?	1
Chapter 2: From Disorder to Injury: Mapping the Terrain(s)	23
Chapter 3: Hermeneutical Circles and Liberative Praxis	53
Chapter 4: The Reification of the Veteran: Kaleidoscopic Lived Experiences	65
Chapter 5: The Centrality of Community in Moral Injury Support:: Theological and Cultural Studies Analysis	91
Chapter 6: Oppositional Forces: Toward a Counterhegemonic Paradigm for Spiritual Care and Counseling	119
Appendix	149
Bibliography	155
Index	163
About the Author	171

Acknowledgments

I am grateful for such a "great cloud of witnesses," many of whom I cannot name directly. However, this book and my vocational callings would not have been possible without the following people and their enduring relational ties. To my family and friends, whether in Kansas City, Pasadena, Long Beach, or Claremont: thank you for listening, loving, and accompanying me to this point.

The dear friends who made up "Unfiltered" on Kandahar Airfield in 2014: Don, Jeremie, and Andrea, you each provided a normalizing and—how strange to say—enjoyable space in combat, thank you. To Tim, John, Erika: you give me hope in the Army Chaplain Corps, and even the Army overall. May we continue to remain crunchy on issues of justice and inclusion, yet joyfully resolute at the prospect of doing our part. Adam, thank you for reading early drafts, articles, and our various conversations about this work. To the soldiers and family members of the 315th Engineer Battalion, I trust that this work honors our service. Serving as your chaplain was one of the most humbling experiences of my life. May "goodness and mercy follow you all the days of your life." *Fabricamos!* Tom, thank you for pushing me and laughing with me as you beat me at FIFA. Matt, I trust there is enough inefficiency in this book to make you proud.

A strange thing happens in graduate school: professors become colleagues who become friends. Duane and Kathleen, in particular, you guided me, pushed me, critiqued me, Skyped in Afghanistan with me, and genuinely know me and what this book represents. A mere "thank you" seems too shallow for the impression you have left. Similarly, fellow graduate students become friends, become colleagues, but remain dear sojourners in this vocation. Marlene, Katherine, Anna, and John, thank you for sharpening whatever analysis and caregiving competencies that follow. Burke Gerstenschlager, thank you for reading my early prospectus and providing an affirmation that I could *actually* write this. The music of Local Natives, The National, and Cold War Kids became the soundtrack to every season of this project, thank you for continually creating art that encourages me to do the same.

Isaac and Annette: you are profoundly loved and known. Live into the joy that you both radiate. Always remember: you are God's beloved. Finally, to Beth: this is our sacrifice. Thank you. From countless drill weekends, TDY trips, and getting through graduate programs, we always knew this was our calling. I adore you, and my love for you only deepens and becomes more steadfast.

Chapter 1

Are We Still Over *There*?

It was different from the other patrols, United States Army Corporal Lisa Fisher thought. The relentless heat and merciless sun were not new; perhaps this was the only constant during her deployment to southeastern Afghanistan. In full kit ("battle rattle") sweating and sitting outside the small Post Exchange (PX in military parlance) patiently waiting for her battle buddies to purchase Monster energy drinks and countless cans of chewing tobacco, she wearily picks up the latest copy of The *Army Times*. The headline, "Woman in Combat: Not in Our Army," struck her at her core. Here she was, one of only fifty women in a brigade combat team of around 4,000 service members. After a deep exhale, she thought: I don't belong here.

After that patrol, during the arduous—yet therapeutic—process of cleaning weapons, her battle buddy continued to reflect on the experience of killing a child. He had no choice, he recalled, as the child emerged from around a corner with a suicide vest strapped to his chest. A silence fell over their scrubbing. What sort of war is this, they thought? He continued to detail for Lisa that he had a daughter about the same age as the child he shot. How will he go home and hold her? Will he ever forgive himself—let alone forget this memory? With each wipe of the cleaning rag, each of them hoping the next wipe would erase their guilt, rage, and the unremitting numbness that accompanied them. Lisa recalls this memory with hesitation. It is a hesitation born out of the scar tissue in her memory: many of her war memories contain the trauma of children. During her time in a Combat Support Hospital (CSH) she witnessed and cared for children the Taliban sent through Unexploded Ordnance (UXO) fields. When she thinks about the mutilated bodies, she notes that the "Taliban killed that child when they put the vest on him." This is a lament—tinged with rage—rather than just relaying fact.

The haunting reality of what follows is how does one come *home* from that? This book is an attempt to honor and lift up those stories. Critically, though, it is a book that asks broader questions of society. There are numerous military moral injury texts, and this book fits within that trajectory, yet there are unraveling narratives which society must hear. As Lisa came home her grandfather commented: "something in you died over there. I can see it in your eyes." Through her eyes, though, she sees civilians thank and applaud her; yet, it takes years for her to truly feel home. She jots in her deployment journal:

> I can't believe it's over—this active duty thing. They give you stamps and papers that are "so important" and mandatory briefings and then the next day you're gone—thrust back into a civilian world. But it's not really over yet inside of you. You're still a Soldier. In the way you move and talk, in your memories and nightmares and flashbacks. It took them years to make me a Soldier. . . . How long will it take me to untie the bootlaces and knots inside of me?

We must all be held to account for Lisa's singular "me" pronoun: this process of untying the bootlaces is also a communal process society must hold in tension.

PATRIOTISM REVISITED

One challenge returning to civilian life from overseas combat is fully reintegrating into a society uninterested in (or at least unwilling to acknowledge) the full psychospiritual impact of war. This is further complicated when that experience includes moral injury—those decisions made in combat that betray an individuals' person understanding of right and wrong—or, what this book will classify as a morally injurious event (MIE). It is my contention that at a societal level, civilians can exacerbate MIEs and re-traumatize returning veterans through our collective inability to reckon with and process society's role in war (with particular attention paid to Iraq and Afghanistan).

This societal exacerbation can be as innocuous as the transactional statement, "Thank you for your service," said when encountering a veteran at the airport, or it can be as extreme as the ideological use of veterans as political pawns to further one's position in national debates.[1] The ongoing controversy over professional athletes choosing to kneel during the national anthem in protest against the brutality of US police forces on Black, indigenous, people of color (BIPOC) communities highlights this ideological extreme. Beginning in 2016 with former San Francisco 49ers quarterback Colin Kaepernick's protest, the kneeling was decoded as disrespectful to service members who "died for the flag." This ideological logic suggests that because veterans died for this country, you *must* stand during the national anthem. The obliviousness of this ideology is seldom taken into account; for example, those who oppose kneeling rarely consider that individuals who "died for the flag" were not afforded adequate access to US Department of Veterans Affairs (VA) health benefits.

The discourse around the anthem expanded from the NFL and has impacted sporting landscapes throughout the world. Protest has forced individuals to confront—or be confronted with—police brutality in a space that is supposedly "just about sports." After the initial COVID-19 shutdown, when sports slowly reemerged throughout the world, moments of silence and raised fists in support for Black Lives Matter reignited a seemingly dormant debate. Arguing that these athletes were *not* protesting military service members and that the players have been clear in what they are protesting is ineffective because the argument takes place within ideology, and as one of my primary interlocutors, Stuart Hall reminds us, ideology is "always contradictory. . . . Ideology works best by suturing lines of argument and emotional investments."[2] Although the kneeling does not deal explicitly with moral injury, it illustrates how parameters of discourse for discussing veterans can affect how returning veterans are received. When ideology sets the discourse for how a veteran is received, a certain *type* of veteran is accepted.

Further, the dominant ideology of mythologizing the military service member maintains a fantasy that any conflict in which the United States is engaged *is* moral. The service members, therefore, who fight these moral wars, are reduced to either "heroes" or "head cases."[3] One dominant ideological narrative of US military service, as pastoral theologian and former military chaplain Zachary Moon posits it, "honors military service on a superficial level and cannot easily accommodate evidence of PTSD, moral injury, and veteran suicide because these phenomena seem to diminish the stoic warrior image."[4] This "stoic" image upholds the hero narrative. On one hand, the veteran is a hero who served one's country with distinction, and the only acceptable relationship to the veteran is one of reverence; on the other hand, as a head case, the veteran is one more broken, substance-abusing, suicidal vet. This binary creates a gap or tension for individuals who have experienced MIEs and are not comfortable seeing themselves as "heroes" or "head cases." Merely reifying a veteran as either a hero or a head case is not tenable as it does not interrogate the conditions of possibility that *created* the situation, and it will only *continue* the current ideological system. Therefore, I argue that without critiquing the dominant ideological system that continues to subject service members to brutal combat rotations, ever-growing suicide rates, and the traumatic experiences of MIEs, society will continue to only see veterans as objects.

Current military moral injury research falls into two strands: one strand focuses on moral injury originating at the individual level and the other strand on moral injury originating at the military organizational level. In terms of the first strand, clinician-researchers Brett Litz, William Nash, and Shira Maguen emphasize the subjective experience of moral injury that occurs when an individual's personal moral code of "what's right" is violated in a high-stakes environment. Litz and colleagues define these experiences as "morally injurious events such as the perpetrating, failing to prevent, or bearing witness to acts that transgress deeply held moral beliefs and expectations."[5] These "deeply" held moral beliefs are "personal and shared familial, cultural, societal, and legal rules for social behavior, either tacit or explicit. Morals are fundamental assumptions about how thing should work and how one should behave in the world."[6]

The organizational level is addressed in the work of former VA psychiatrist Jonathan Shay. Shay focuses on a betrayal of "what's right" by a significant authority figure in a leadership position, which is not so much an individual moral injury, but a moral injury produced through incompetent (or failed) leadership. For example, Shay opens his groundbreaking text, *Achilles in Vietnam: Combat Trauma and the Undoing of Character*, with a vignette of US service members in Vietnam.[7] After firing upon three boats that American forces suspected were used to transport weapons, the service members realized that fishermen and children occupied the boats. The "confusion" set in for the service members when leadership told them, "It is fine," and "Don't worry about it."[8] The commanding officer told them it was "fine," that leadership would "take care of it," to which one service member responded, "So you know in your heart it's wrong, but at the time, here's your superior telling you that it was okay."[9]

The two strands of moral injury research are not mutually exclusive but interdependent, as they coexist. One strand can lead into the other. A MIE that starts in the organizational context can easily coalesce into an individual MIE as a soldier fails to do "what's right" after key leadership individuals have failed to uphold "what's right." To a degree, these concepts are helpful in differentiating moral injury from posttraumatic stress disorder (PTSD). However, something is missed in typologies that isolate the event in either the individual or the organizational construct of the military. Scholar Joseph Wiinikka-Lydon also traces the ambiguous genealogy of moral injury, noting that "clinicians lose their critique, as their understanding of moral injury focuses more narrowly on the individual soldier rather than on the political and institutional context in which they were injured."[10] I think Wiinikka-Lydon is right, and this project is centered on the systemic and political implications of returning from combat.

Further, a focus on individual pathology does not critique the broader social and cultural implications of reintegrating with the experience of MIEs, which would call into question current foreign policy and the ideological (mis)use of military service for political and social purposes. As caregivers, military chaplains need to be able to conceptualize the ideological apparatuses at work in MIEs in order to find a way through their matrices to cocreate meaning and move toward healing.[11] Otherwise, as suggested by pastoral theologian Bruce Rogers-Vaughn, caregivers are "condemned to become chaplains to the status quo and particularly on behalf of any hegemony that happens to be in power."[12] That warning is further cemented as Uruguayan Jesuit priest Juan Luis Segundo, one of the founders of Latin American liberation theology, reminds us, "We must understand and appreciate the ideological mechanisms of established society if theology is to take the word of God and convert it from a vague outline to a clearly worked out message. Otherwise theology will become and remain the unwitting spokesman of the experiences and ideas of the ruling factions and classes."[13]

Beyond the battlefield, chaplains seeking to reintegrate veterans into civil society need a clearer conceptualization of cultural ideologies that can amplify an existing moral injury. Therefore, I examine how American society's ideological mythologizing of military service functions and how it affects the reintegrating veterans who have experienced moral injury by using a qualitative case study methodology in conjunction with methods found within a liberative praxis approach to practical and pastoral theology.

In reimagining moral injury, I seek to add my own nuance, which is that reintegrating with a moral injury is further complicated within a cultural ecosystem steeped in an ideology of the significance and mythologizing of military service members. Thus, my focus is to provide an ideological and societal critique of how people with military moral injuries are viewed and treated. I assert that to properly reintegrate veterans into society, two steps are necessary: First, veterans need a space in which to fully articulate and "re-author" their experiences of MIEs.[14] Second, military chaplains (primarily) and clinicians (secondarily) must critique the dominant ideologies of American military service to cocreate counterhegemonic spaces of proper reintegration. A cocreated space is one of solidarity, as military chaplains position themselves as "organic intellectuals." Italian Marxist Antonio Gramsci strategized that for a counterhegemonic group to attain hegemony, organic intellectuals from the same social class are needed to organize and lead from within that class (more on this in chapters 2 and 6). Military chaplains can realize such an intellectual role. If ideological critique of the reintegration of veterans with moral injuries is not properly utilized, veterans will continue to receive the hollow platitude of "thank you for your service," while nothing will change at a cultural, political, and theological level. At worst, the United States' service members will continue to die in the wars that have become our longest, and the wars will continue to wreak havoc on generations of communities throughout the world.

Theologian Joseph McDonald posits that the level of personal agency is what differentiates the two types of military moral injury.[15] This is a helpful analysis, and it assists in conceptualizing the phenomenon of moral injury. However, emphasizing agency as the site of moral injury research is problematic for two key reasons. First, it falls into an objective, pathologizing medical worldview that privileges diagnosis. Diagnosis establishes a hierarchy of "objective" truth in which, in a traditional counseling scenario, power resides with the therapist or other professional providing the diagnosis. This has had success, of course, but it does not contribute to long-term reintegration because it fails to reintegrate the veteran back into a society that is not prepared, ideologically, to conceptualize moral injury. In other words, the veteran has done "the work," in therapy, but the society is not prepared to reintegrate the veteran back into it.

Anecdotally, some veterans who come to see me for spiritual care and counseling in my role as an army chaplain are reticent to go to the VA in fear of the aforementioned stigma of being considered a "head case." Some research supports this anecdote, as there is a reported inverse relationship between rates of self-reporting behavioral and mental health concerns and rates of accessing treatment facilities.[16] Admitting a psychological need can still be perceived as a symptom of failure or a sign of weakness. Two of the veterans I interviewed herein agree. The accompanying guilt, which is a primary symptom of moral injury, "often prevents many veterans from seeking social and spiritual support. Without such support, veterans will remain stuck in guilt, and less able to co-create more complex meanings about morally injurious events."[17] These individuals, then, fall through the cracks; in avoiding the perception of failure, they do not access resources that could benefit their reintegration.

Inherently, the issue becomes one of the ideology of the military as a system that upholds "duty," "selfless service," and "honor," which can seem untenable with self-reporting a behavioral or mental health concern. It is not merely an issue of self-reporting, as Roseanne Visco unpacks in her research with active duty Air Force airmen: airmen were more likely to self-report physical symptoms than behavioral and mental symptoms.[18] The ideological malaise surrounding behavioral and mental health symptoms is precisely why military chaplains need to be able to decipher the ideological matrices; military chaplains represent a more "socially accepted source of help" than "traditional mental health providers."[19]

Within the military, chaplains are trusted in part because they are present in the lives of their service members, living on the "front line with regard to MI [moral injury] counseling."[20] Chaplains are *where* the service members are; they *live* as the service members live; and they *suffer* as the service members suffer. Chaplains in many different contexts understand this as a ministry of presence. Indiana University religious studies professor Winnifred Fallers Sullivan unpacks the multivalence of a ministry of presence as follows: "In some contexts for some people, presence can be reassuringly immanent and down to earth, empty of formal doctrinal content, comfortingly abstracted from tradition, but in and for others, it can specifically evoke highly elaborated theological understandings and rituals."[21] Another critical reason why military chaplains are, perhaps, preferable to more traditional behavioral health counselors is the absolute confidentiality a chaplain offers. A service member knows that when he or she visits a chaplain "there is no appointment officially recorded . . . that would otherwise show up on medical documents or other records."[22]

Second, and crucially, focusing on agency fails to account for the societal and structural implications of MIEs. More specifically, focusing on agency fails to take into consideration how American military service functions as an ideology in ways that can exacerbate the reintegration struggles of veterans with moral injury. Caregivers need to embrace both a spiritual care paradigm that emphasizes a communal response while also critiquing how ideology functions in society. As pastoral theologians have emphasized since at least Bonnie Miller-McLemore's 1993 article on the "living human web," the caregiving relationship includes a broader social and political consciousness of how webs of interdependence impact persons.[23]

With her concept of the living human web, Miller-McLemore voiced concerns with Anton Boisen's method of studying the "living human document."[24] Boisen's living human document method posited that the same rigor that went into studying biblical texts or literary texts could be applied to individuals in a caregiving relationship. However, Miller-McLemore's primary conversation partners of liberation theologies and feminism elucidated how systems contribute to domination and oppression. To take seriously the concerns of marginalized communities, spiritual care and counseling needed imagery that spoke to those concerns. Miller-McLemore's shift, then, was to move away from a narrowly defined understanding of counseling to a broader, more inclusive focus on the contextual aspects of care. Her conceptual contribution to the field of pastoral theology focused on the embeddedness of persons in various public webs of meaning.

Miller-McLemore is not advocating the complete erasure of the living human document as an image by which to understand care; rather, she has a broader conceptualization of care that takes the holistic approach of taking into account the multifaceted systemic realities at play. "Ultimately," she says, "I suggest that 'the living human document within the web' is the metaphor that best captures the subject matter of both CPE and pastoral theology."[25] With this trajectory of pastoral theology behind and surrounding me, I contend that focusing on individual agency without considering societal responsibility will not lead to a holistic reintegration for veterans *or* society.

Pastoral theologian Ryan LaMothe has effectively argued that the American hegemonic narrative of "superiority, exceptionalism, and innocence" has successfully obfuscated an ability to grieve properly, and I would add to this that this narrative shapes how we mythologize our veterans.[26] Dominant ideologies impact the ways in which we receive veterans and the ways in which we understand combat trauma—and even military service broadly speaking.

I advocate for developing a clearer theoretical conceptualization of cultural ideologies that can amplify an existing moral injury. In developing this conceptualization, I turn to several theological, political, philosophical, and psychological resources. I meld the communal-contextual spiritual care paradigm with a liberative praxis spiritual care practice of solidarity. Liberative praxis, which is found within a methodological strand of hermeneutics in practical and pastoral theology privileges the marginalized and oppressed in communities of faith and works to liberate these individuals to various pathways of freedom. Liberative praxis relies on theological insights from liberation theologies—such as Latin American, Black, Latinx, Queer, Womanist, and Feminist—including God's preferential option for (and solidarity with) the marginalized, a praxis-based ecclesiology, and a God who is experienced within history (and who even experiences suffering). These insights deepen the care interventions. For example, Jessica Vazquez-Torres utilizes moral injury as a way to understand racialization and particularly how frameworks of racial structures prevent the potential solidarity for those on the margins. In effect, it becomes "seen exclusively as a struggle between black and white, which elides the experience and struggles of other communities of color."[27] Liberative praxis starts from lived experience and embodies solidarity.

All this to say: this is a book built on and committed to solidarity. Solidarity is more than empathy; it is the recognition that not only are we all in this together, but also how, then, can we act to change our circumstances. "Thank you for your service" is empathy, helpful, yet not committed to changing the material experiences of veterans and civilians. A commitment to solidarity is more than a minor theological and political entrenchment. There is the potential for tension and pushback: where does this book fit: pro-war, anti-war? However, what these binary categories miss, or easily gloss over, are the ways in which we are all complicit. When we take a pro-war stance, out of reverence for the veterans that serve in the military, we ostensibly neglect the glaring injustices done by systems. If we take an anti-war stance, out of various political or theological perspectives, we ostensibly neglect the individual lives caught up in the cogs of the machine.

Pastoral theologian Sharon Thornton aptly articulates solidarity as "being willing to stay beside people in their struggle to be free from unnecessary suffering, being willing to fight with them and not necessarily for them for their release."[28] That fight is the spirit of what undergirds this project. I seek to develop this solidarity through an ecclesiology that will support the reintegration of veterans by drawing from an analysis of Dietrich Bonhoeffer's *Sanctorum Communio: A Theological Study of the Sociology of the Church* (*Sanctorum Communio: Eine dogmatische Untersuchung zur Soziologie der Kirche*); and insights from ideology critique developed by cultural studies theorist Stuart Hall and significant adaptation and reimagining Gramsci's counterhegemonic proposals.

BRIEF CONTOURS AND ROAD MAPS

In chapter 2 I will provide a thorough map of the terrains that hold this project together, so here I will briefly note the primary trajectories and the through line from the aforementioned conversation partners that will enhance this book. First, as this is a project on moral injury, I review the first generation of moral injury research, tracing its development from Jonathan Shay through the handful of pastoral theologians addressing the phenomenon of moral injury. The other important strand to unpack is ideology critique, in particular my primary interlocutors of cultural studies and critical theory, Hall and Gramsci.

Moral Injury

As the literature on moral injury is emerging from its first generation, most of the research is descriptive in nature. Therefore, much of early moral injury research has consisted of differentiating the phenomenon from PTSD. With PTSD only entering the *Diagnostic and Statistical Manual of Mental Disorders* (DSM) as a diagnosis in 1980, the production of not only moral injury research but also PTSD research is somewhat recent. However, this makes it sound as though there has been a clear demarcation between the diagnosis of PTSD and of moral injury, and this is not the case. Jonathan Shay's work with Vietnam veterans is considered the first to differentiate the stressors that lead to moral injury from PTSD criterion.

Shay's work with Vietnam veterans decisively found that for an MIE to occur, three things all need to happen: (1) there has to be a betrayal of what is morally right, (2) by someone who holds legitimate authority, and (3) in a high-stakes situation.[29] As I mentioned above, in 2009, Litz and his colleagues isolated the MIE in the subjective experience of individual veterans. They understood morally injurious events to include "the perpetrating, failing to prevent, or bearing witness to acts that transgress deeply held moral beliefs and expectations."[30] In the time between Shay's description of the phenomenon of moral injury in 1994 to Litz and his colleague's transitional definition in 2009, not much changed. Even after 2009, the psychological literature continued to build primarily on these two definitions.

Within theological studies broadly, moral injury first gained attention through Rita Nakashima Brock and Gabriella Lettini's *Soul Repair: Recovering from Moral Injury after War*.[31] In addition to cowriting this text, Brock was the first codirector of the Soul Repair Center at Brite Divinity School, Texas Christian University, in Fort Worth, Texas. Brock and her team of veterans and chaplains traveled the United States providing seminars and workshops to people interested in moral injury.

Ideology Critique

I am turning to cultural studies and ideology critique to correct the psychological focus on individual suffering and pathology that dominates the literature of moral injury. Although the burgeoning literature of moral injury, for the most part, originated in the field of psychology, it cannot remain there. Ideology critique offers assistance with unraveling a more complete reckoning, and the communal practices I advocate enacting are discussed in chapters 5 and 6.

Therefore, although chapter 2 unpacks the technical understanding of ideology that originated with Karl Marx and was expanded upon by Hall, I want to provide here this project's working definition of ideology. It is merely *working* because, as Terry Eagleton reminds the reader "nobody has yet to come up with a single adequate definition of ideology."[32] For this project, an *ideology* is the set of ideas by which people structure their existence and rationalize inconsistencies within their worldviews. An ideology is, therefore, any set of beliefs that determines social behavior. Building on this concept, ideology critique traditionally sought to point out the inconsistencies of beliefs in the system and to show people (or "unmask" for them) that their ideological beliefs do not reflect reality as it "really" is. The goal, at least in the lineage from Marx's ideology critique, is to get people away from their ideological beliefs and into an unideological reality. The critique of ideology becomes a mechanism that "presumes that nobody is ever *wholly* mystified—that those subject to oppression experience *even now* hopes and desires which could only be realistically fulfilled by a transformation of their material conditions."[33] The tension remains, however, in whether this more psychoanalytical perspective of critique is realistic: something is needed beyond unraveling peoples' poor political analysis of their subjugation. Another tradition of critique is available.

I utilize the work of Stuart Hall, the second director of the Centre for Contemporary Cultural Studies (CCCS) at the University of Birmingham from 1968 to 1979. It is Hall's ability to synthesize, critique, and move past his conversation partners that I contend makes his theorizing of ideology more complete than others. His theorizing on ideology is more complete via his understanding of the media *as* ideological. For Hall, the world is saturated with media, and humanity understands and produces meaning through media.

Hall rejects the binary distinction between high culture and popular culture, and he consistently notes that politics resides at the level of popular culture. For these reasons, Hall was just as comfortable exegeting British soap operas, mining them for meaning, as he was explicating Marx or Gramsci. As Hall himself stated, "Indeed, if we put the emphasis on the mass or multiple production of cultural goods, we must include all books, newspapers, gramophone records, etc., Tolstoy as well as Spillane, The *Times* as well as the *Daily Mirror*, Beethoven as well as the Beatles."[34]

The media, those artifacts of popular culture, are crucial for Hall's work because they become instantiations for how people understand themselves. In other words, meaning is produced in popular culture, and this production is inscribed onto media products. In the veteran support discourse, the utilization of media is of critical relevance. I will specifically dissect films and various cable news representations of veteran news stories as they relate to my veteran participants' experiences. Returning to Hall, ideology is not a false consciousness, but rather, socially constructed. There is no "one to one relationship between the conditions of social existence we are living and how we experience them."[35] Ideology has an existing logic to it that creates and limits social and metaphysical understandings of how the world works. Also, there is no *outside* of ideology. For Hall, "it is not possible to bring ideology to an end and simply live the real."[36] That idea itself is, simply, pure ideology. The challenge, or opportunity, is about *which* ideological frameworks we engage, rather than how to escape them.

METHODOLOGY AND METHODS

Just as I briefly outlined the closely related literature above, I will summarize my methodology here before providing a complete discussion of my methodology and methods of care in chapter 3. At this point, I want to note the methodological goal of this practical theology project. The goal is twofold. First, I seek to identify how, through dominant ideologies, society has failed to reintegrate returning veterans. Second, I seek to provide an account of how veterans can privilege their MIEs as sources of authority and wisdom in talking to others about war and reintegration, without positioning themselves as either "heroes" or "head cases." There are three methods that guide my approach: (1) a practical theological hermeneutical method of liberative praxis, (2) the qualitative methodology of case study, and (3) the cultural studies approach of ideology critique.

The qualitative inquiry method for this study is case study. I am using specific examples of four veterans to consider the broader experience of reintegration with a moral injury. I am committed to exploring the interdependent webs that Miller-McLemore raised. What is it like to return home after combat? This is a question for all returning service members, but there are unique circumstances informing the narratives herein that need investigating. For two of the veterans, it is unique, for example, that Reserve component service members can possibly return to their civilian work setting or educational setting only weeks after participating in an overseas combat deployment. Once these individuals "redeploy" to the United States, there is a set amount of time during which service members are evaluated and screened for potential traumatic exposure or other physical, mental, and emotional ailments.

However, most service members will report whatever is deemed necessary to get home and potentially fail to disclose an MIE, therefore missing the opportunity to arrange follow-up counseling. Once these service members get home, behavioral health resources are not as readily available for them. This could simply be an issue of proximity, as their home could be miles away from not only a military installation—or a VA center—but also miles away from a community that understands military service. This community does not necessarily need to be made up of veterans, but the community needs to have familiarity with veteran support.

I interviewed four combat veterans of Operation Iraqi Freedom (OIF) and/or Operation Enduring Freedom (OEF): Army Reserve Specialist (SPC) Phillip Campbell, Army Corporal (CPL) Lisa Fisher, Army Reserve Private First Class (PFC) Angela Gallagher, and U.S. Army Sergeant (SGT) Andrew Lloyd.[37] In chapter 4 I will more formally introduce their stories. It was critical to also interview and analyze veterans from the Reserve component of the military. The U.S. military is divided into components: active duty and the Reserve component. The National Guard (Army and Air Force) and the Reserves (Army, Air Force, Navy, and Marine Corps) make up what is referred to as the "Reserve component." For the Reserve component, military service is part-time, in contrast with the full-time active duty forces. Due to this part-time status, there is a presumption that Reserve component service members are less professional or that they are "weekend warriors who only know about a war zone from the latest iteration of *Call of Duty*."[38] Reserve component service typically includes a monthly "battle assembly," more colloquially referred to as "drill." In theory, the Reserve component service members accomplish a month's worth of tasks in one weekend. During the summer months, Reserve component service members attend various annual training events throughout the country, typically over a two-week period.

Once service members are deployed, there is no difference between the type of mission in which the Regular Army and the Reserve component participates. In the wars in Iraq and Afghanistan, an estimated 46 percent of service members have come from the Reserve component.

In 2007, the US Secretary of Defense released a memorandum outlining the deployment cycle for the Reserve component. For every year an active duty member was deployed to a combat zone, he or she was to receive two years dwell time at home station (to train and prepare for the next rotation, once again, in theory). A Reserve component member was to receive five years of nondeployed time for every year of deployment.[39] However, this has not been a reality for either component. The Congressional Budget Office has reported that even when a unit is stateside, its members are not actually at home station but are equipping and training to deploy again.[40] Further, when deployed, the deployment of "boots on ground" has extended to fifteen months (with some deployments even extended to eighteen months).

AUDIENCE

This work fills a gap in intervention by moving from a solely psychological approach to a treatment plan that is an interdisciplinary practical and pastoral theology. First, I am writing to professionals in the field of spiritual care, specifically military chaplains, local religious leaders, and clinicians, namely VA chaplains. Military chaplains are often limited in direct support of veterans and typically interact with a veteran for a short window of time; long-term care happens in VA facilities or faith communities throughout the country. Because of this continuity of care, military chaplains *and* caregivers in a civilian context need to conceptualize what is at play within the veteran space. It is not enough to provide individualistic measures of resiliency and ritual; unless the entire system is called into question, we would be fools to think anything will change.

Second, I am writing to concerned civilians. It is my belief through anecdotal conversations that there is a concerned civilian population that wants to find ways to support the veterans in their community. They, too, understand that our community as a whole—military and civilian—is bound up in the discourse surrounding war. Third, as this work is firmly located within post-Marxist literature, I am writing to scholars who see inroads with their scholarship on post-Marxists analysis: namely work that aims at dismantling neoliberal capitalist hegemony, white supremacy, and heteronormativity. It is my argument that a holistic approach to an anti-war stance would bring together scholars with similar theological and political commitments.

COMING HOME: FROM DISILLUSIONMENT TO REIMAGINATION

This is inherently a book about coming home. In many ways, though, it is a book about claiming one's voice, and hopefully in those ways, it becomes a fuller story of my own coming home. As I embark on trying to tell *a* war story, perhaps I am coming to know more about who I am on the other side of my combat deployment. Within this ongoing process of coming to know oneself my hope remains one in which:

> To know oneself means to be oneself, to be master of oneself, to distinguish oneself, to free oneself from a state of chaos, to exist as an element of order but of one's own order and one's own discipline in striving for an ideal. And we cannot be successful in this unless we know others, their history, the successive efforts they have made to be what they are . . . and we must learn all this without losing sight of the ultimate aim: to know oneself better through others and to know others better through oneself.[41]

With that in mind, I want to name my locatedness as this relates to the project. I am an ordained minister in the United Church of Christ (UCC) as well as an United States Army Reserve chaplain. I will speak from my tradition while honoring the multivalent backgrounds and traditions of the participants in this book, just as I do with *all* my soldiers. My situatedness as a "progressive" Christian chaplain influenced by liberation theologies entails that I find beauty, truth, meaning, and complexity in every religious tradition.

I received my commission as an officer in the United States Army Reserve and was immediately placed in the Chaplain Candidate Program in 2008. Prior to my commissioning, to think that I would wear the uniform was not necessarily on my radar. It was never that I was anti-military. My father and grandfathers served in the military, but their service was at a time before I was an idea. Perhaps the idea never entered my mind, because in Kansas City, where I was born and raised, the military was not a part of my psyche. My journey toward the chaplain corps, like any good story, began with rejection. In 2006 I had this idea: I was going to be an attorney; well, more specifically, a public defender—a modern-day Atticus Finch fighting for the cause of justice. The only problem is that my dream forgot to inform the law school admissions committees. So, with multiple rejection letters in hand I changed course and moved to Europe and spent a year ministering to military teens whose parents were stationed on US military installations in Germany. Throughout that year I witnessed families living through the trauma of the surge in Iraq. I witnessed families receive the news that their loved one was killed in action (KIA). I left Germany with the conviction that providing care to this community was paramount. This story, and others, start to unravel my own interdependences and places in which I can trace my theological commitments and political alignments around veterans issues.

During my army chaplain career, I have served at various echelons both stateside and one combat deployment, in 2014, to Afghanistan. I left Afghanistan at the end of 2014, and left what I thought was the end of Operation Enduring Freedom (OEF). I thought the wars in Iraq and Afghanistan were over. I was proud that I was a part of its conclusion. Coming home, though, was somewhat my moment of disillusionment. There was no single, isolated, incident; rather, it was the ongoing frustration that nothing changed—and nothing would change. As combat operations continued, a season of disillusionment followed. While completing my doctoral program in Southern California, my own political commitments began to shift. I have supported the Democratic Party for my entire, yet admittedly brief, adult life. My commitments have ranged from directly supporting, and volunteering for, the John Kerry presidential campaign in 2004 as a College Democrat at the University of Missouri–Kansas City to a role in which voting was my sole civic duty.

My disillusionment grew for the two primary US political parties: the Democratic Party and the Republican Party have not put forward a consistent and realistic plan to support the ending of the wars in Iraq and Afghanistan. Nor have they offered a holistic plan to care for veterans. The December 2019 revelations from the *Washington Post* in their "Afghanistan Papers" only revealed this fact. The *Post*, perhaps trying to match their journalistic endeavors reached with releasing Daniel Ellsberg's documents that comprised the "Pentagon Papers" in 1971, probably wondered why the 2019 version of leaked government documents landed with a whimper and quickly left the national discourse. However, when I read the "Afghanistan Papers" I read that three bipartisan presidential administrations "were devoid of a fundamental understanding of Afghanistan—we didn't know what we were doing."[42]

I want to be clear: I was not necessarily surprised by those revelations as the political process, as it currently functions, does not adequately have the interests of veterans in mind. The burgeoning moral injury research contained in this book assisted me in coming to terms with my realization that I felt betrayed by the military, and by my country. The "Afghanistan Papers" are the rule of these wars, not the exception. Perhaps my surprise came with how quickly the passion and outrage extinguished. Yet, we are still fighting. We are still dying. These political implication are morally injurious, and I seek to address these critiques directly. Wiinikka-Lydon states this succinctly: "Moral injury, then, is a visceral experience of policy, as well as cultural assumptions, that are put into effect corporately on the ground through the bodies of soldiers and others."[43]

After my deployment I continued to provide spiritual care and counseling to soldiers and their families after multiple combat tours. Marriages and familial relationships were rattled, frayed, and destroyed by nearly two decades of US war efforts. From my perspective as a caregiver, my frustrations grew: these wars were supposed to be over. What I needed, then, is precisely what this book hopes to provide: a critical and liberative analysis of moral injury, specifically how an American ideology of military service can exacerbate veterans' moral injuries and how counterhegemonic groups can offer a reimagined way forward.

ORIGINALITY AND CONTRIBUTIONS

Is there a need for another moral injury text? First-generation moral injury texts have, with precision and clarity, defined the phenomena. There is a space for this book, though as it contributes to the fields of practical theology and pastoral theology in four ways. First, I am integrating moral injury and ideology critique. The focus on individual pathology has not adequately critiqued the broader social and cultural implications of MIEs that call military service into question and the dominant cultural ideologies about the military. Focusing on ideology compels society to examine, in close detail, how our treatment of veterans (through our practices or beliefs about military service) shapes us as a society.

Chaplains need to be able to conceptualize the ideological apparatuses at work in order to find a way through those matrices. Further, although there are abundant Marxian critiques of economics, neoliberalism, and intersectionality within spiritual care and counseling, there is a lack of post-Marxist ideology critique in the field. I bring these conversation partners into closer proximity and more explicitly discuss how ideology functions and ways in which spiritual care providers can move through the malaise.

Second, I am adding the societal role in moral injury as an explicit influence in the reintegration of people with traumatic experiences. Building on my first point, the role of society is vastly underutilized. Veterans are not isolated individuals; rather, veterans come from and return to communities of people. Countless moral injury resources note that healing happens within communities; therefore, what role does society plays in reintegration? The inherent military and civilian divide does not help this reintegration either. This book will uniquely show via interviews and other research the prominence of society's role in exacerbating moral injury.

Third, I am exploring how moral codes and identity are reconstructed at the communal level. This further contributes to research that privileges the role of community over and against research that remains pathologizing and focused on diagnosis. Finally, military chaplains are traditionally understood as "force multipliers." Historically, the US military chaplaincies have functioned to provide religious support to assigned service members. Within the function of offering religious support, a chaplain might provide the religious rites and services he or she is authorized to perform as an ordained or credentialed religious leader. However, particularly in the post-9/11 landscape, military chaplains are now tasked with new functions that include advising the command on issues of morale, morals, ethics, and religion and *how* these factors impact the current mission. In combat especially, the military expects its chaplains to rally behind the cause of the country. Theologian Ed Waggoner, drawing from Pentagon documents, states that chaplains "as a multiplier of force . . . speak martially, patriotically, and divinely all at once."[44] I argue that military chaplains can function in an additional way that supports praxiological commitments of solidarity by functioning as Gramscian intellectuals. Chaplains, even within the hegemonic military-industrial complex, can provide solidaristic support to veterans with moral injury.

OUTLINE OF THE CHAPTERS

Chapter 2 picks up from the complexity of the malaise within chapter 1 with a genealogy of pertinent literatures within the discourses of moral injury, ideology critique, and practical theology. These threads present the kaleidoscopic framework which provides not only the robust critique of the project, but also offers a *new* paradigm of care and counseling. The paradigm presented in chapters 5 and 6 receives its scaffolding in chapter 2. While seemingly spinning multiple plates, chapter 2 unpacks the congruence and through line of these intersecting discourses.

Chapter 3 explains the liberative praxis model of care building upon the work of practical and pastoral theologian Emmanuel Lartey. Chapter 4 is narrative-driven and analyzes the experience of what was learned from the interaction between the veteran participants and myself. I will detail three prominent themes of alienation: belonging, divided identities, and betrayal. Within that betrayal, there is a theological component as well: Phillip and Lisa, in particular, felt abandoned by God and distanced from their primary religious resources of prayer, community accountability, and support. It is also important to note, that even with the case study of the two participants, Angela and Andrew, who do not identify as Christian, the goal is that the liturgy of solidarity in chapter 6 would still provide supportive structures for them.

Chapter 5 summarizes an understanding of God's pathos from liberation theologies and how this is vital for understanding the reintegration of veterans with MIEs. Interacting with Hall's concept of ideology, this chapter attempts to decipher whether it is possible to exist within the ideological apparatuses present in US society. This chapter argues that Hall's "oppositional" reading of decoded texts is possible, paving the way for a more holistic communal reintegration. Bonhoeffer upholds this communal reintegration.

Employing a close study of Bonhoeffer's early work, specifically his doctoral dissertation, which was later published as *Sanctorum Communio: A Theological Study of the Sociology of the Church* and his posthumously published *Letters and Papers from Prison*, this chapter demonstrates that he upholds this communal reintegration.[45] These works enhance and support a theological understanding of the importance of community and a God that is present within the traumatic experiences of an MIE. Bonhoeffer's insights are operationalized through the communal-contextual paradigm of spiritual care and counseling.

Finally, chapter 6 provides a summary of this project and offers paradigmatic proposals for military chaplains who provide care to military veterans. This chapter also recommends how veterans can privilege their MIEs in opposing the United States' ideological mythologizing of military service. Furthermore, it argues that chaplains can empower veterans to resist the marginalization of their experiences and that care requires social advocacy through public discourse. Part of this empowerment is to assist veterans negotiate the tensions inherent within newfound freedoms. As a caregiver, the goal of the chaplain is not to romanticize counterhegemony, but to grapple with the risks of opposing dominant ideologies. Building from Hall's insights in chapter 5, Antonio Gramsci's "war of position" offers a way to understand the formation of counterhegemonic groups. Military chaplains function as Gramscian intellectuals and live in conscious solidarity with their service members. The specific spiritual practices of conscious solidarity complete my practical theological method. These practices include a liturgy of solidarity through which a revitalized prayer life and the sharing of war stories is cultivated in a counterhegemonic community. The inclusion of both communal practices *and* individual practices upholds my commitment to envisioning a solidarity-based pastoral theology that cultivates action rather than platitudes.

NOTES

1. Joshua T. Morris, "'Thank You for Your Service:' Mapping Counter-Memories as a Form of Spiritual Care Support for Moral Injury," *Journal of Pastoral Theology* 28, no. 1 (2018): 34–44.
2. Stuart Hall, "The Neoliberal Revolution," in *Selected Political Writings: The Great Moving Right Show and Other Essays*, ed. Sally Davison, David Featherstone,

Michael Rustin, and Bill Schwarz (Durham, NC: Duke University Press, 2017), 326.
3. Zachary Moon, "Turn Now, My Vindication is at Stake:' Military Moral Injury and Communities of Faith," *Pastoral Psychology* 68, no. 1 (February 2019): 94; *Warriors Between Worlds: Moral Injury and Identities in Crisis* (Lanham: Rowman & Littlefield Publishing Group, 2019), 84–86.
4. Moon, *Warriors Between Worlds*, 84.
5. Brett T. Litz et al., "Moral Injury and Moral Repair in War Veterans: A Preliminary Model and Intervention Strategy," *Clinical Psychology Review* 29, no. 8 (2009): 697.
6. Litz et al., "Moral Injury," 699, quoted in Warren Kinghorn, "Combat Trauma and Moral Fragmentation: A Theological Account of Moral Injury," *Journal of the Society of Christian Ethics* 32, no. 2 (2012): 60–61.
7. Jonathan Shay, *Achilles in Vietnam: Combat Trauma and the Undoing of Character* (New York: Scribner, 1994).
8. Ibid., 4.
9. Ibid.
10. Joseph Wiinikka-Lydon, "Mapping Moral Injury: Comparing Discourses of Moral Harm," *Journal of Medicine and Philosophy* 44, no. 2 (April, 2019): 179.
11. In chapter 2 I will discuss with much greater depth how I am utilizing "apparatus." For now, I am operationalizing it as French structuralist and Marxist, Louis Althusser detailed.
12. Bruce Rogers-Vaughn, "Best Practices in Pastoral Counseling: Is Theology Necessary?" *Journal of Pastoral Theology* 23, no. 1 (2013): 4.
13. Juan Luis Segundo, *Liberation of Theology*, trans. John Drury (Eugene, OR: Wipf & Stock Publishers, 2002), 39.
14. Michael White, *Maps of Narrative Practice* (New York: W.W. Norton & Company, 2007), 61.
15. Joseph McDonald, introduction to *Exploring Moral Injury in Sacred Texts*, ed. Joseph McDonald (London: Jessica Kingsley Publishers, 2017), 14.
16. Roseanne Visco, "Postdeployment, Self-Reporting of Mental Health Problems, and Barriers to Care," *Perspectives in Psychiatric Care; Madison* 45, no. 4 (October 2009): 247.
17. Carrie Doehring, "Military Moral Injury: An Evidence-Based and Intercultural Approach to Spiritual Care," *Pastoral Psychology* 68, no. 1 (February 2019): 16.
18. Visco, "Postdeployment," 241.
19. J. Irene Harris, et al., "Moral Injury and Psycho-Spiritual Development: Considering the Developmental Context," *Spirituality in Clinical Practice* 2 (January 1, 2015): 257.
20. Lindsay B. Carey et al., "Moral Injury, Spiritual Care and the Role of Chaplains: An Exploratory Scoping Review of Literature and Resources," *Journal of Religion and Health* 55, no. 4 (2016): 1230.
21. Winnifred Fallers Sullivan, *A Ministry of Presence: Chaplaincy, Spiritual Care, and the Law* (Chicago: The University of Chicago Press, 2014), 174.
22. Carey et al., "Moral Injury," 1230.

23. Bonnie J. Miller-McLemore, "The Human Web: Reflections on the State of Pastoral Theology," *Christian Century* 110, no. 11 (April 7, 1993): 366–369.
24. Anton T. Boisen, *The Exploration of the Inner World: A Study of Mental Disorder and Religious Experience* (Philadelphia: University of Pennsylvania Press, 1936), 10.
25. Bonnie J. Miller-McLemore, *Christian Theology in Practice: Discovering a Discipline* (Grand Rapids, MI: William B. Eerdmans Publishing Company, 2012), 51.
26. Ryan LaMothe, "Empire, Systemic Violence, and the Refusal to Mourn: A Pastoral Political Perspective," *Journal of Pastoral Theology* 23, no. 2 (2013): 2.
27. Jessica Vazquez-Torres, "Does Moral Injury Have a Color? On Moral Injury and Race in the United States." Paper presented at the annual meeting of the American Academy of Religion, 2014, quoted in Wiinikka-Lydon, "Mapping Moral Injury," 186.
28. Sharon G. Thornton, *Broken Yet Beloved: A Pastoral Theology of the Cross* (St. Louis: Chalice Press, 2002), 123.
29. Jonathan Shay, "Moral Injury," *Psychoanalytic Psychology* 31, no. 2 (2014): 183.
30. Litz et al., "Moral Injury," 697.
31. Rita Nakashima Brock and Gabriella Lettini, *Soul Repair: Recovering from Moral Injury after War* (Boston: Beacon Press, 2012).
32. Terry Eagleton, *Ideology: An Introduction* (London: Verso, 2007), 1.
33. Ibid., xiv.
34. Stuart Hall and Paddy Whannel, *The Popular Arts* (Durham, NC: Duke University Press, 2016), 35.
35. Stuart Hall, "Signification, Representation, Ideology: Althusser and the Post-Structuralist Debates," *Critical Studies in Mass Communications* 2 (1985): 105.
36. Stuart Hall, "Ideology and Ideological Struggle," in *Cultural Studies 1983: A Theoretical History*, ed. Jennifer Daryl Slack and Lawrence Grossberg (Durham, NC: Duke University Press, 2016), 138.
37. All names have been changed to protect privacy. My participants were given the option to pick a name, have the principal investigator pick a name, or decide to use their name.
38. Joseph Kassabian, *The Hooligans of Kandahar: Not All War Stories Are Heroic* (TCK Publishing, 2017), location 267, Kindle.
39. Terri Tanielian and Lisa H. Jaycox, *Invisible Wounds of War: Psychological and Cognitive Injuries, Their Consequences, and Services to Assist Recovery* (Santa Monica, CA: RAND Corporation, 2008), 23.
40. Ibid.
41. Antonio Gramsci, "Socialism and Culture," in *The Antonio Gramsci Reader: Selected Political Writings 1916–1935*, ed. David Forgacs (New York: New York University Press, 2000), 59.
42. Craig Whitlock et al., "A Secret History of the War," *Washington Post*, December 9, 2019, accessed July 21, 2020.

43. Joseph Wiinikka-Lydon, "Moral Injury as Inherent Political Critique: The Prophetic Possibilities of a New Term," *Political Theology* 18, no. 3 (2017): 228.
44. Edward Waggoner, "Taking Religion Seriously in the U.S. Military: The Chaplaincy as a National Strategic Asset," *Journal of the American Academy of Religion* 82, no. 3 (September 2014): 718.
45. Dietrich Bonhoeffer, *Sanctorum Communio: A Theological Study of the Sociology of the Church*, Dietrich Bonhoeffer Works, vol. 1, ed. Clifford J. Green, trans. Joachim Von Soosten, Reinhard Kraus, and Nancy Lukens (Minneapolis: Fortress Press, 2009); Dietrich Bonhoeffer, *Letters and Papers from Prison*, ed. Eberhard Bethge (New York: Simon & Schuster, 1997).

Chapter 2

From Disorder to Injury

Mapping the Terrain(s)

As this project is distinctly a work of practical theology, I meld multiple academic disciplines to better conceptualize the ideological aspects of reintegrating people with moral injuries into civilian life. This genealogy, then, is an attempt to map these influences together to enhance the analysis of the lived experience of my veteran participants (see chapter 4). First, through the discipline of practical theology, I am interested in its liberative hermeneutical work. Second, I detail what I am defining as the "first generation" of moral injury research, tracing its development from Jonathan Shay and other clinician-researchers, while also noting the handful of theologians interacting with the concept of moral injury. In the third and final section, I excavate a primary cognate conversational partner of British cultural studies, in particular the post-Marxist ideology critique of theorist Stuart Hall. I turn to ideology critique to correct the silo effect of the psychological focus on individual suffering and pathology that dominates moral injury literature. To properly situate Hall's contribution, I plot a course of ideology critique from Karl Marx through two subsequent Marxists more conducive to Hall's project: Antonio Gramsci and Louis Althusser.

PRACTICAL THEOLOGY

In this section, I narrate practical theology's story in two parts. First, I address the field's ambiguity through American practical theologian Bonnie Miller-McLemore, who offers four helpful usages of the term *practical theology* and their relevance for the field. Second, in conjunction with the history of Miller-McLemore's third interpretation of *practical theology*—"an approach to theology and religious faith"—I focus on the renaissance of practical theology in the twentieth century. This renaissance foreshadows the liberative praxis further developed in chapter 3.

Four Uses of Practical Theology

Bonnie J. Miller-McLemore states that four things are indicated by the term *practical theology*: "an activity of believers," a "curricular area," an "approach to theology and religious faith," and an "academic discipline."[1] In the first usage of the term—an activity of believers—Miller-McLemore is interested in naming the connections made by "believers" between their beliefs and their practices. Second, she identifies the usage of the term for the program of study in seminaries and divinity schools that is the curricular area called "practical theology." Practical theology as a curricular area is of note as the subdisciplines are (typically) pastoral care, homiletics, mission, evangelism, leadership, and education. These fields are connected under the auspices of practical theology because their common goal is to enrich ministerial practice.

The third usage of the term practical theology—an approach to theology and religious faith—places practical theology as a method. This is primarily how this project operationalizes practical theology: I, as a practical theologian, am seeking to critically understand how veterans in different locations and under different circumstances make meaning out of MIEs and then attempt to integrate these experiences into civilian life. To say this is to hold in tension a field that has moved from application to a theology that has "pragmatic" and "normative" tasks.[2] Miller-McLemore identifies this as a focus on *telos*, in that practical theologians are working toward a larger end. The *telos* of this project is learning from my veteran participants how ideology impacts reintegration. From that understanding and lived experience(s), I proceed to focus on intervention and spiritual care practices to support veterans.

Finally, Miller-McLemore asserts that *practical theology* is used to refer to an academic discipline, what Gijsbert Dingemans calls a "science of action (*Handlungswissenschaft*)."[3] Dingemans is a helpful conversation partner here because he names what enables the discipline to thrive: its "empirical-analytical," "hermeneutical," and "critical-political approaches."[4] The empirical-analytical approach collects descriptive facts about a community. The hermeneutical approach moves away from the quantitative focus of the empirical-analytical aspect and creates a "thick description" of a community and its context.[5] The critical-political approach begins by focusing on the oppressed in communities of faith and then brings their concerns back to the academy, the church, and the society. This approach works in tandem with conversation partners such as Latin American liberation theologians, Black liberation theologians, and Feminist and Womanist theologians. The concerns of the oppressed have helped forge the renaissance of the discipline of practical theology.

Practical Theology's Renaissance

In the wake of the failure of the liberal experiment of human progress in the face of the catastrophic destruction, dehumanization, and death caused by the First and Second World Wars, theology needed a substantive reevaluation of its mission. Practical theology sought new inroads for relevance in this postmodern cultural context. Practical theologians pushed the boundaries of research and discovered new areas of contribution. They were beginning to ask fresh questions, and these questions were no longer questions solely about epistemology or morality; rather, the questions in the mid-twentieth century centered on the subject matter, the tasks, and the theological nature of practical theology.

This shift, of course, did not occur in a vacuum; the rise of post-Christendom and practical philosophy—what American practical theologian Andrew Root describes as "pragmatism, Marxism, and Aristotelian perspectives"—highlighted an elevated view of reflexivity.[6] The elevated view of reflexivity made the case that the living human document is as credible a source of truth as other texts. Anton Boisen, the founder of clinical pastoral education (CPE), developed the living human document method, positing that the same rigor that went into studying biblical texts or literary texts could be applied to individuals in a caregiving relationship. Practical theology had new conversation partners as well, including psychology and the ethnographic advances in the social sciences (e.g., anthropology and sociology). Emerging medical models of training and supervision that emphasized clinical placement and case studies transformed practical theological reflection.

Latin American liberation theologies, Black liberation theologies, Feminist and Womanist theologies gained increasing visibility. These voices prioritized emerging pedagogies, epistemologies, and practices from the oppressed to challenge traditional religious practices and doctrines. In an incisive comment on the importance of liberation theologies, Elaine Graham, Heather Watson, and Frances Ward state that oppressed communities worked to "'*democratize*' *theology* as a 'work of the people,' in an effort to return it to those on the 'underside of history' whose voices and perspectives were formerly neglected."[7] The democratization of theology within traditional Christian denominations is more than a minor footnote in practical theology's story. This empowering of the laity ushered in paradigms in which a theory of action was implemented in the world. It changed who gets to define theology.

All of this change is not without critique, however. Feminist practical theologian Rebecca Chopp cautions against allowing these new methodologies, particularly the liberationist paradigm, and the renaissance overall from becoming just a veiled attempt at maintaining modernity's (and theology's) status quo. Her critique specifically concerns how the correlation methods of liberal revisionism are operationalized in such a way that they continue to oppress the oppressed by taking agency from them.[8] Her critique is critical for this project: namely, how tenable is it to speak a word of liberation from within one of history's most hegemonic forces (i.e., the military-industrial-complex)? The usage of cultural studies assists in answering those concerns, while liberative praxis continues to be vital for calling systems of oppression into question, even from the periphery.

As a guidepost for what is to come in chapter 3, I want to conclude with praxis. My primary clinical commitments are of solidarity manifesting through praxis. Praxis is not merely a reflection on action, but a reflection on action *and* reflection that inherently works to influence and change both. Praxis critically examines the fields of study as I am moving the conversation beyond descriptive and interpretive understandings of moral injury toward a critical analysis of how practices are enacted that handle ideology and its place in social life.

To truly understand praxis from a Marxist perspective, one must understand that it is never merely a reflection on practice; it must involve change, or "revolution." This happens through a dialectic of practice, reflection on practice, and the synthesis of praxis. The material interests of the ruling class drive society, and therefore, these interests determine how social relations function. To be emancipated from this situation, the goal is not to engage abstractly with the ruling class but to struggle for change through an elevated self-consciousness. In other words, the goal is an *orthopraxis* of changed behavior rather than an *orthodoxy* of right belief. Therefore, as Marx himself famously stated, "The philosophers have only interpreted the world in different ways; the point is to change it."[9] With Chopp's warning (and Marx's maxim) in mind, the goal is to change how we understand ideology in order to better reintegrate veterans. Thus, a more concentrated review of moral injury research is warranted.

MORAL INJURY

I want to situate moral injury in its historical context. First, I will note Jonathan Shay's work with Vietnam veterans as a historical precedent for the current work on moral injury. After Shay's initial work, first-generation moral injury research has shifted to focus on individual acts of commission or omission, and these insights are fully ensconced within psychology. Second, I will highlight the various theologians interacting with moral injury and discuss their work to develop interventions for care. I conclude this section with a global critique of these theologians (particularly the pastoral theologians) and their work on moral injury: namely, they seek to enact practices that do not critique the ideological myths of military service. Without critiquing dominant ideologies, American veterans will continue to experience a gap in treatment; therefore, the binary of "hero" or "head case" prevails.

From Disorder to Injury: Moral Injury's History

It is prudent to begin with Jonathan Shay, as his work with Vietnam veterans is considered the first to differentiate the stressors leading to moral injury from the criterion of PTSD.[10] Further, Shay moved away from referring to these injuries as "disorders" and instead called them "injuries." He writes, "Combat PTSD, is a war injury. Veterans with Combat PTSD are war wounded, carrying the burdens of sacrifice for the rest of us as surely as the amputees, the burned, the blind, and the paralyzed carry them."[11] Shay's work noted three necessary categories to classify an event as morally injurious: (1) there has been a betrayal of what is morally right, (2) by someone who holds legitimate authority, and (3) in a high-stakes situation.[12] Shay's work has also been well received because he connects the experience of betrayal in Vietnam with the Homeric narrative. Shay describes this connection as an "experiment" that reveals "living knowledge to us today."[13] For example, Shay describes how, within Homer's epic poem *The Iliad*, Achilles experiences a betrayal (i.e., a moral injury) by his commander, Agamemnon, and how Achilles responds to this betrayal. Shay's goal is to attempt to underscore a sort of universality of moral ambiguities inherently prevalent in combat. To elaborate,

> The epics teach no lesson at all to modern forces on weapons, planning, communications, tactics, organization, training, or logistics. But for those who go to war and return from it today, the epics still vibrate with meaning on cohesion, leadership, and ethics.[14]

I agree with Shay that there are principles of combat that have remained unchanged, particularly in the sense of what it means to kill, to watch others get killed, and to attempt to integrate this into any sense of normalcy at home.

However, Shay's conflation and universalizing of the experiences of *all* wars misses several essential nuances of the wars in Iraq and Afghanistan. Primarily, in the operating environment of the post-9/11 Global War on Terrorism, war is asymmetrical. A symmetrical war is a traditional conflict between states of roughly equal personnel strength and weapons capabilities.

In contrast, an asymmetrical war is fought unconventionally, without traditional nation-state backing, and it is fought on multimodal platforms in what Canadian military chaplain Steve Moore helpfully dissects as "intrastate" warfare.[15] To elaborate, the Islamic State in Iraq and Syria (ISIS, ISIL, or IS) is able to recruit from outside Iraq and Syria because it is not a bounded state (although they seek to establish an Islamic Caliphate in the region around Iraq and Syria—the Levant). Individuals are able to pledge allegiance to ISIS without being in Iraq or Syria. This means that when terrorist attacks happen across the world—away from the Levant—ISIS is able to claim responsibility for them. This is asymmetrical warfare. Homer discusses the calamities of war, but it is symmetrical war he addresses; collapsing all warfare into a generalized category is not helpful.

In 2009, Brett Litz and his clinical psychology colleagues expanded the definition of moral injury beyond betrayal by authority figures to isolate the MIE in the subjective agential experience of individual veterans. Their moral injury became "the perpetrating, failing to prevent, or bearing witness to acts that transgress deeply held moral beliefs and expectations."[16] In the space between when Shay noted the phenomenon of moral injury in the early 1990s and the clinical work from Litz and his colleagues, not much changed, even as the literature burgeoned. Therefore, following 2009, the military moral injury discourse continued to build primarily on these two definitions.

There is an additional thread, away from the military discourse that is imperative to discuss. This thread falls outside the scope of the book; however, it does provide some scaffolding for care considerations. Joseph Wiinikka-Lydon differentiates the moral injury literature into three separate locales: the clinical, the juridical-critical, and the structural.[17] The literature I have just unpacked is located within the clinical discourse; namely, psychology and psychiatry. The juridical-critical framework developed out of philosophy and philosophy of law. Within that philosophical tradition, critical theorist Axel Honneth[18] and J. M. Bernstein approach moral injury as an injury of misrecognition. With respect to Bernstein, he unpacks moral injury and its dehumanizing violence in acts of rape and torture.[19] Wiinikka-Lydon describes this theoretical move as having "no special context of institution in the juridical discourse, such as combat zones or the military. Instead, moral injury can result from various contexts, from daily social interactions to extreme cases, such as rape, torture, or hate crimes."[20]

Finally, Wiinikka-Lydon distinguishes moral injury within structures. This discourse brings the reader full circle to Shay; yet, the structural discourse pushes the bounds of what Shay addresses. The structural model is able to hold the both/and tension of moral injury: yes, there are agents who commit acts of moral injury, but there are also structural impediments in place that perpetuate moral injury, such as race. The late practical theologian Dale P.

Andrews notes after reading Rita Nakashima Brock and Gabriella Lettini's *Soul Repair: Recovering from Moral Injury after War* that the anti-racism discourse is ripe for examination.[21] For Andrews, the examination centers on the white dominant culture's hesitancy, or resistance, to take moral responsibility for racism. The lack of responsibility is not a conservative or liberal issue: racism, and the lack of redressing its implications, is the betrayal of core moral beliefs. Racism, in Wiinikka-Lydon's estimation "stunt[s] the growth of empathy."[22]

Returning to the clinical framework, the definitions provided by Shay and Litz were reified, even in Shay's own writing, with "Moral Injury N for Nash, Litz, & Maguen" referring to the individual definition (the "N" designating Nash) and "Moral Injury S, for Shay" standing for Shay's authority-figure definition of moral injury.[23] Shay has been quick to note that Moral Injury N is still a viable and important strand of moral injury research; he just does not believe he can *proactively* do anything about it, short of "ending the human practice of war."[24] Shay currently works with military leaders and highlights the importance of unit cohesion, ethical leadership, and realistic combat training simulation to prepare individuals for Moral Injury S events. It is somewhat easier to prepare leaders not to fail morally than to solve issues inherent in warfare, such as the implications of killing noncombatants (i.e., women and children).

I have proceeded thus far as if military communities as a whole—or even all research clinicians—accept that moral injury should be included alongside PTSD as a combat trauma phenomenon. This, however, is not the case. Even naming the phenomenon "moral" and "injury" is problematic for some. Injury implies physiological harm or damage, so some have offered "moral affront," "moral distress," "moral conflict," "moral pain," "moral trauma," "moral wounds," "moral disruption,"[25] "combat-activated onto-ecological disorientation,"[26] and "inner conflict" as alternative terms.[27] The Navy and Marine Corps's doctrinal publication *Combat and Operational Stress Control* insists on designating "stress arising due to moral damage from carrying out or bearing witness to acts or failures to act that violate deeply held belief systems" as *inner conflict* instead of *moral injury*, since *moral injury* was "perceived by some to be pejorative."[28] Following this pattern, others have difficulties with the *moral* part of *moral injury*, preferring instead "emotional injury," "personal values injury," "life values injury," and "spiritual injury."[29]

Further still, there is an element in the literature, particularly as it relates to interventions, that holds that the *experience* of something as morally injurious points to a moral identity that is working properly, in that "moral injury cannot afflict a sociopath."[30] In other words, if an individual did not experience a traumatic response from betraying the individual's moral identity, then a separate phenomenon is possibly taking place. However, this level of pathologizing is not entirely helpful. A categorization in which *not* experiencing an MIE as morally injurious is considered pathological is something that will further isolate the experience of individual veterans and might work to alienate them.

Not surprisingly then, within this ambiguity and minor reification, there is no universally agreed-upon classification of moral injury. Some researchers note that due to moral injury's empirical infancy, it should continue to fall under the auspices of PTSD. Others, including Shay, note the similarities but are resolute in differentiating the two experiences; according to Kent D. Drescher and colleagues, there is "universal agreement among subject matter experts that the concept of 'moral injury' is needed."[31]

I join Shay in emphasizing that moral injury, unlike PTSD, is not primarily connected to neurological responses of fight or flight ("physiological arousal"), and PTSD does not "necessarily involve shame and guilt."[32] Affectively, moral injury is usually a shame response (among other responses listed below) based on actions perpetrated or witnessed in combat. There are common symptoms identified in moral injury, and some of these overlap with PTSD. First, moral injury and PTSD are similar in symptomology in that anger, depression, anxiety, insomnia, and nightmares are common to both, and these symptoms are often self-medicated via substance abuse. But to lump the two phenomena together—although there are overlaps—is to do a disservice to each one.

Moral injury and PTSD are distinct in how they present in an individual, and to merely recommend the same treatment intervention for both is its own morally injurious event. For example, prolonged exposure (PE) and cognitive behavioral therapy (CBT) are the primary intervention models for PTSD groups within the VA, and neither of these paradigms is necessarily helpful for an individual with moral injury. First, in PE, the goal is to address a specific memory by recounting the memory to effectively overcome the stressor through desensitizing oneself to the memory. PE encourages participants to face their fear, so to speak, by confronting a traumatic memory. CBT attempts to change a participant's thoughts about a behavior or experience, because, in this paradigm, if one changes one's thoughts, one can possibly change one's actions. However, these therapeutic formats that rely on reexperiencing and reliving an event are not necessarily beneficial and are possibly injurious for an individual with a complex moral injury.

Clinical psychologist Jeremy Jinkerson has attempted to develop a symptomology for MIEs, which will effectively assist in conceptualizing a better communal paradigm for MIE interventions. The core symptomology of moral injury includes guilt, shame, spiritual or existential conflict (including "subjective loss of meaning in life") and a "loss of trust in self, others, and a transcendent or divine entity."[33] At a secondary level, Jinkerson notes the overlapping features I mentioned above: depression, anxiety, anger, reexperiencing the moral conflict, self-harm, and social problems.[34] Although the psychological community is not ready to designate moral injury as a diagnosable condition, Jinkerson still notes that "for moral injury to be identified, the following criteria must be present: (a) a history of morally injurious event exposure, (b) guilt, and (c) at least two additional symptoms, which may be from either the core or secondary symptomatic feature lists."[35]

Within a discussion on symptomology, there is further ambiguity with respect to what becomes a moral injury. What happens in the space between an MIE and a moral injury? In Shay's work, he notes how the military organizational construct can lead to an MIE (whether through the incompetence of leadership or through a lack of accountability). The operating environment of combat can also serve as the platform for an MIE. The intrastate conflicts of asymmetrical battlefields present the ambiguity of threats and personnel. Systematic theologian Duane Larson and chaplain Jeff Zust offer an equation that speaks to this dynamic phenomenon. They state,

> One could almost propose a mathematical equation that displays how these terms and events relate, wherein MIE represents morally injurious event, MP is moral pain, and MD refers to moral dissonance. These are all discrete matters in the literature. Thus MIE + MP MD. And if moral dissonance is left unaddressed, and the damage compounds, then MD MI. Together these events and pain are elements of a destructive dissonance that results in moral injury.[36]

How have clinician-researchers recommended intervening with these destructive dissonances? Within the literature, there is a phenomenon in which the intended audiences reside in silos. There is space for an interdisciplinary role, but there are few sources collaboratively working together to treat MIEs.[37] For example, psychological essays are "colonized" for psychologists.[38]

As an army reserve chaplain, I can gain insights by reading Jinkerson's work or that of Litz and his colleagues, but their works provide limited insights for spiritual care. To elaborate, in *Adaptive Disclosure: A New Treatment for Military Trauma, Loss, and Moral Injury*, Litz and his colleagues offer a practical guide for practitioners on how to work with service members on their journey through moral injury.[39] Adaptive disclosure (AD) is a six-session treatment process created for service members coming to terms with three types of traumatic war experience: life threat, loss, and moral injury. AD is unique in that it is a psychotherapeutic intervention geared solely toward war trauma. An entire chapter of the text is devoted to teaching nonmilitary therapists about the military cultural context in order for them to provide acute-level psychotherapy. Litz details the assumptions of AD in another article:

> The following assumptions guide our approach: (a) pain means hope. Guilt and shame (from perpetration-based moral injury) and anger (from betrayal-based moral injury) are signs of an intact conscience and expectations of the self and others about goodness, humanity, and justice; (b) goodness is reclaimable; and (c) forgiveness (when applicable, feasible, and therapeutically valuable) and repair are possible.[40]

However, within AD's strict CBT-based interventions, a role for religion and spirituality is absent, though religious and spiritual beliefs can offer crucial therapeutic interventions and meaning-making for some service members. The importance of a "benevolent moral authority," confession, meaning-making, and forgiveness are mentioned in *Adaptive Disclosure* and other places within the Litz canon, but spiritual intervention methods are missing.[41] Therefore, assumption (c) above is offered as a guide without a mechanism to provide forgiveness and restoration. Beyond the importance of differentiating moral injury from PTSD, as a chaplain, Litz's text limits my professional scope in how to treat veterans experiencing moral injury. What is ultimately missed in AD is a thorough analysis of the spiritual implications of moral injury.

In 2018, however, Litz began to acknowledge the role of spirituality in interventions. This acknowledgment led him to alter his intervention and give it a new designator, "adaptive disclosure-enhanced" (AD-E). AD-E is now a twelve-session intervention that appropriates Buddhist spiritual practices of loving-kindness meditation (LKM) to "break through rigidity, numbness, hopelessness, and disconnection, and in the case of moral injury caused by others, anger and resentment (and potential revenge fantasies)."[42] This is a positive step forward, and once randomized controlled trials are implemented, a better understanding of LKM's usage can be proffered.

Within the AD-E intervention, LKM is used as "compassion training."[43] Service members are trained to become aware of their own suffering and the suffering of others and to develop an intention of kindness and compassion to self and other through meditation. I wholeheartedly support veterans utilizing meditative practices, and I think it might do wonders to counteract a military culture in which meditation is seemingly anathema; however, this training is never done with a Buddhist practitioner or with a community of fellow practitioners. The AD-E therapist has the latitude to decide how much and how often to utilize LKM. My critique of AD-E is that appropriating LKM detached from a community of practitioners colonizes a spiritual practice of a specific sect of Buddhism. Further, there is a necessity to avoid an intervention that "unwittingly serves the neoliberal hegemony, the very system which is intensifying and multiplying the sorts of sufferings that bring people to psychotherapists."[44]

It is too early to tell how successful AD-E will be; however, despite the critiques, utilizing LKM is a step in the right direction as it utilizes spirituality and enacts practices that are "designed to bolster service members' and veterans' sense of shared humanity and connection to others. It reduces the distance between oneself and others."[45]

One intervention model that focuses on the spiritual distress of combat is Irene Harris' Building Spiritual Strengths (BSS). BSS is an interfaith intervention model that addresses the spiritual distress dimension of MIEs and emphasizes the existing spiritual resources to make meaning of an MIE. BSS has shown that "those who view their spirituality, faith community, and/or Higher Power as sources of support, validation, and acceptance are more able to make healthy meaning and recover than those who don't."[46] Unlike AD and AD-E, BSS is an eight-session group-based intervention in which people "(a) establish group rules and develop rapport, (b) use a modified empty-chair technique to facilitate dialogue with a Higher power or similar spiritual construct, (c) explore prayer/meditative coping techniques, (d) explore theodicy (spiritual explanations for suffering), (e) explore and reframe forgiveness of self and others, and (f) plan for continued support for spiritual growth."[47] BSS is an optimal intervention for reintegration as it takes the communal aspect of trauma—and reintegrating that trauma—seriously. I will discuss the importance of relying on techniques (c), (d), and (e) in chapter 6's liturgy of solidarity.

Spiritual Implications for Intervention

Just as Shay and Litz and his colleagues are indispensable to first-generation moral injury research, womanist theologian Rita Nakashima Brock and feminist theologian Gabriella Lettini are as well. They have been able to achieve two things that have brought theological awareness to moral injury. First, in December 2008, Brock and Lettini began working alongside more than sixty religious leaders, activists, veterans groups, and academics in the Truth Commission on Conscience in War to advocate further education on moral injury. From their work came the 2012 text *Soul Repair: Recovering from Moral Injury after War*, which offered an entryway into moral injury research for many chaplains, ministers, and laypeople concerned with issues surrounding military involvement. Through familial narratives and firsthand accounts of reintegration, Brock and Lettini not only highlight the concept of moral injury but also propose theological insights into communal restoration methods that solidify recovery and treatment within religious communities.

Second, Brock was the first codirector of the Soul Repair Center at Brite Divinity School, Texas Christian University, Fort Worth, Texas. While in that role, Brock and her team of veterans and chaplains traveled the United States providing seminars and workshops to people interested in moral injury. Brock has since left that role to pastoral theologian Nancy Ramsay, and at present, Brock is the Senior Vice President for Moral Injury Programs at Volunteers of America. My primary critique of Brock and Lettini, however, is more global, and it ties in with theological accounts of moral injury. I will provide my critique after describing the work of additional pastoral theologians.

The late Larry Graham's *Moral Injury: Restoring Wounded Souls* identifies how a person's understanding of the transcendent impacts not only the view of self, but also how it compounds moral conundrums that can lead to moral injuries. In some events, God *is* the moral conundrum, as is the case in some of the upcoming vignettes. A veteran might ask, "How could God allow this event to happen? Where was God?" Graham explores alternative ways of naming and framing relational theologies (such as process theology, liberation theologies, and feminist theologies) in which God is present in solidarity with the sufferer. Graham is clear that healing is possible, even when one is dealing with the classical theodicy conundrum.

Graham's pastoral theological work has consistently sought a relational justice that will integrate people *into* communities and will create understanding that people come *from* communities. These insights from Graham offer an excellent entry point into my concern: namely, the communal piece ultimately fails because of the systemic functioning of ideology. Graham describes the "enculturated" systems of ideology as "macrosystemic moral gyroscopes."[48] He gets to the precipice of communal involvement and reintegration, but backs away to instead focus on "collaborative conversations" that empower "anyone" to normalize the experience of moral injury; however, these collaborative conversations do not engage ideology.[49]

Within those collaborative conversations, he also attempts to broaden what we understand as moral injury. Graham pushes the boundaries of moral injury as a concept past combat and military moral injury. Graham uses the framework of combat moral injury to generalize it for everyday use in that "we all struggle with moral injury, when moral injury is broadly understood as the failure to live in accordance with our deepest moral aspirations."[50] Graham, however, keeps the aforementioned moral injury framework in his schema: moral injury is still either received (receptive) or given (agential).

There is an implicit cognitive dissonance in Graham's work in that he can maintain that macrosystemic moral gyroscopes impact individuals and communities; however, his practices are ones of intra- and interpersonal healing. Graham can state that "much of the moral dissonance, dilemmas, and injury we face today takes its rise because of irreconcilable differences within the macrosystemic moral orientations into which we are enculturated," yet the practices he describes are interpersonal.[51] Both pastoral theologian Carrie Doehring and I have separately noted threads in Graham's work on the politics of lamentation that are perfectly situated to provide an intervention for moral injury.[52]

Zachary Moon, pastoral theologian and former military chaplain, offers practical steps for local clergy and congregation members to think about how to reintegrate returning veterans. His work is essential for the provision of competent spiritual care within the current military structure. Moon's first monograph, *Coming Home: Ministry that Matters with Veterans and Military Families*, lays some responsibility for this reintegration on the doorstep of the church, and he aptly provides resources for those who are looking for practical ways to support veterans.[53]

Moon's major contributions to the field of moral injury research are his understanding of the limits of our binary view of veterans as "heroes" and "head cases" and his structural analysis concerning how military basic training prepares recruits to adopt the military's "moral orienting system." As he states succinctly,

> Military recruit training, by design, destabilizes and diminishes the constancy of a recruit's pre-existing moral orienting system. Having stripped away such moral coding, including embedded values, beliefs, behaviors, and meaningful relationships, military recruit training indoctrinates recruits with a new moral orienting system that supports functioning in military contexts.[54]

Moon's analysis is that the betrayal of what is right in a combat situation is encapsulated within the broader military cultural construct. Recruits are enculturated into that construct once their previous way of operating is deconstructed in basic training. In combat, cohesive units are vital for survival. However, it is at home, during reintegration, when "irreconcilable dissonances between one's moral orienting systems and the moral worlds one inhabits" collide.[55]

IMPLICATIONS FOR PRACTICAL THEOLOGY AND THIS PROJECT

My primary critique of these pastoral and practical theologians as they relate to the broader practical theological project on moral injury is that their work is *within* the dominant ideological structure that I am saying is necessary to critique. Remaining within dominant ideologies fundamentally complicates reintegration: while offering a critical differentiation of moral injury from

PTSD and enacting spiritual practices, pastoral theologians writing about moral injury have seemingly left the existing dominant ideological structure unaddressed. Ultimately, dominant ideologies are left free from critique, and, left untested, dominant ideologies will continue to bolster support for endless cycles of combat deployments, which can lead to further MIEs.

For example, in Graham's *Moral Injury*, he discusses the story of Gary, an Afghanistan combat veteran. Gary's story fits within Graham's broader section on healing collaborations, so the aim of Gary's vignette is to elucidate how caregivers can provide support to those suffering from moral injuries. Gary is introduced to a pastoral counselor, Ellen. Ellen and Gary have a vulnerable and powerful caregiving relationship, one built on curiosity, awareness of each one's emotions, and a cocreative process of healing. Graham concludes the section by stating, "There must be a phase of revising one's self-assessment and moral actions in light of these prior steps."[56] At this point, there is a distance between the caregiver and the care receiver. Gary was with Ellen in cocreating a healing narrative; however, in this stage, Gary is alone in a process of re-evaluating his involvement in war that is "very difficult, and takes moral courage too."[57] I would argue that Gary is alone during one of the most precarious points of his reintegration. A process is needed in which both Gary and Ellen revise their moral actions in relation to war. What is Ellen curious about? Where are the places in which Gary's story has caused Ellen to think differently about war, moral injury, and reintegration? Finally, where are the places in which Ellen is healed in the cocreated narrative? To make the final assessment of and action on the MIE solely the responsibility of the veteran is to ultimately fail both the veteran and the caregiver at a crucial point.

What is needed, then, is a critical look at the ideological systems and how they reproduce dominant ideologies. What I am proposing is envisioning how ideology functions to exacerbate existing MIEs. To envision this exacerbation, we must be willing to challenge the ways in which ideology teaches us to think about military service. To achieve this, cultural studies and ideology critique are my primary theoretical interlocutors.

IDEOLOGY CRITIQUE

I turn to ideology critique to correct the silo effect of the psychological focus on individual suffering and pathology that dominates moral injury literature. Remember, ideology is "necessary for an understanding of interests."[58] Building on this, ideology critique, beginning with Marx, has sought to point out the inconsistencies of such beliefs in the dominant system and to show (or "unmask") to people that ideological beliefs do not reflect reality as it "really" is. The goal is to get people away from ideological beliefs and into an unideological reality. What follows is a tracing of this technical strand of ideology critique originating with Marx.

Karl Marx

By way of introduction to Marx's approach to ideology, I want to start with his conception of culture. In *The German Ideology*, Karl Marx and Friedrich Engels posit that "as individuals express their life, so they are. What they are, therefore, coincides with their production, both with *what* they produce and with *how* they produce."[59] Further, they go to great lengths to argue that culture is found within humanity's "double relation" to nature and humanity.[60] Humanity utilizes and manipulates nature in order to reproduce material life. Naturally, according to Marx and Engels, the manipulation of nature moves into a social organization (whether that be exchanging goods or labor) to more effectively reproduce material life. They argue that this social organization is apparent throughout history.

However, where it connects to ideology is through their materialist view of history. For Marx and Engels, there is no labor in general, but rather, labor and production are anchored in history. They state, "the fact is, therefore, that definite individuals who are productively active in a definite way enter into these definite social and political relations."[61] Hall helpfully reiterates that Marx's concept of culture refers to "the arrangement—the *forms*—assumed by social existence under determinate historical conditions."[62]

In terms of this historical materialism, Marx and Engels begin to map how ideas, concepts, and consciousness are produced. The difficulty enters when such maps are experienced in a way that does not map onto reality. Ideology, then, is a *false* consciousness, a distortion, and an inversion of reality. When Marx and Engels describe ideology as a *false* consciousness it should not be read to claim either that it does not exist or that it is repressed, as in a Freudian framework; rather, it is a false consciousness in that the idea cannot fully encapsulate the entire reality that the idea seeks to describe. Marx and Engels also refer to this phenomenon as a *"camera obscura,"* in that humanity appears "upside-down," like an "inversion of objects on the retina does from their physical life-process."[63] Humans are thrown off from the *actual* conditions of their lives and depend on conditions that they have no control over. These conditions are found within the anatomy of social production: the base and the superstructure.

Ideology is a determining factor within the "base" of Marx's economic theory. Within this theory, there is a base and a superstructure. The base, which contains the means of production (i.e., tools and machines) and relations of production (i.e., commodities), determines the superstructure. The superstructure (e.g., media, culture, or religion) functions ideologically and reflects the values of the ruling class. Culture reflects, or is determined by, the economic base.

In a classic passage from *The German Ideology*, Marx and Engels assert that the ideas of the ruling class always correlate to the positions of the ruling class (what I will henceforth refer to as the Marx-Engels principle). These ideas are located in a specific ideology, or, as they state:

> The ideas of the ruling class are in every epoch the ruling ideas, i.e., the class which is the ruling material force of society, is at the same time its ruling intellectual force. The class which has the means of material production at its disposal, has control at the same time over the means of mental production, so that thereby, generally speaking, the ideas of those who lack the means of mental production are subject to it. The ruling ideas are nothing more than the ideal expression of the dominant material relationships, the dominant material relationships grasped as ideas; hence of the relationships which make the one class the ruling one, therefore, the ideas of its dominance.[64]

Hall addresses whether it is possible to posit a ruling-class ideology, but Hall's overall development builds on critiques from two important Marxist theorists: Louis Althusser and Antonio Gramsci. Gramsci influenced Althusser, but I will begin by discussing Althusser due to the incomparable influence of Gramsci on Hall and Hall's work within the CCCS and the New Left.

Louis Althusser

Althusser broke with traditional Marxism—and therefore the Marx-Engels principle—with his three critiques on ideology. First, for Althusser, it is problematic to assume that the ideological position of a social class will relate directly to its "position in the social relations of production."[65] Second, the premise of false consciousness presumes a certain epistemological relationship. Third, and most important at present, Althusser, in his foundational essay "Ideology and Ideological State Apparatuses (Notes towards an Investigation)," posits that knowledge is produced in practices.[66] Within this critique, Althusser is suggesting that ideological practices are invisible *objectively*. For Althusser, "It is irrelevant whether people are fooled by the ideological lies—ideology sustains its grip either way."[67] Thus, we may say, with Althusser, "ideology is *misrecognized* at the level of people's practices, rather than merely *misrepresented* at the level of their ideas."[68] The "Ideological State Apparatuses" essay is substantial in that it unpacks Althusser's understanding of human subjectivity and explains how people continue to reproduce the conditions that maintain their domination.

The first move in the "Ideological State Apparatuses" essay is to look for the *function* of ideology. Ideology works to reproduce social relations of production. In a capitalist social system, labor is reproduced outside social relations. This is because capitalism is not just a means of production; it also cultivates a need within people to buy the products being produced. How, though, does this ideology function? This moves into Althusser's second thesis: ideology has a material existence. Althusser posits, "An ideology always exists in an apparatus, and its practice, or practices. This existence is material."[69] Hall is helpful here in clarifying that a material existence includes practices and customs that provide "the 'ideas' with which people figure out how the social world works, what their place is in it, and what they *ought* to do."[70] Ideology is inscribed on people's existences. One's identity is acquired through material practices, not through what one believes.

Althusser differentiates between "ideological state apparatuses" and the "repressive state apparatuses." The ideological state apparatuses (ISAs) include religious, educational, familial, legal, political, trade-union, communication (media), and cultural (arts) mechanisms. The repressive state apparatuses (RSAs) include the government, the administration (e.g., taxation), the army, the police, the courts, and the prisons.[71] Therefore, these apparatuses function either through ideology (ideological state apparatuses) or through coercion and violence (repressive state apparatuses). RSAs secure by force the reproduction of production. An example may prove helpful. An employee at a factory cannot take home the product he or she has produced; it is owned by the company, and this is reinforced through the police. The police—or factory security—are called when an individual tries to take a product out of the factory.

In terms of ISAs, Althusser's comments that the school replaced the church as the dominant ideological state apparatus. No other apparatus has the "obligatory (and not least, free) audience" to inculcate an ideology into.[72] Schools teach the ultimate bourgeoisie principles of freedom and responsibility as adults. This ideological state apparatus teaches students the skills that will help them succeed in society. Schools, then, provide an apprenticeship in ideology, that is, on-the-job training in ideology. For Althusser, though, this guarantees that students are subject to the state, and ideology "kicks in" as individuals begin to believe the system because they have already been acting in the system.

The final thesis of the "Ideological State Apparatuses" essay is how ideology constitutes subjects. Borrowing from psychoanalyst Jacques Lacan, ideology "interpellates" individuals as subjects. People are "hailed" (another term for interpellation) by unconscious ideologies that enlist them as subjects. Interpellation is one's response to a police officer shouting, "Hey you, stop!" What does the police officer's call produce within an individual? The response is retrospective of the call. As Althusser states, "By this mere one-hundred-and-eighty-degree physical conversion, he becomes a subject. Why? Because he has recognized that the hail was *'really' him* who was hailed (and not someone else)."[73]

With these theses in mind, a basic Althusserian definition of ideology could be "the imaginary relationship of individuals to their real conditions of existence."[74] Althusser's work bridges the gap to later theorists on ideology (e.g., Slovenian cultural theorist Slavoj Žižek) in that Althusser utilizes Blaise Pascal's argument (found in *Pensées*) that if one wants to believe, one must simply act, and belief will follow. An example may help solidify this idea. An employee believes that working hard could one day benefit the employee, even while the employee becomes fatigued, and the ideology of the system reinforces the work ethic as a "value," while never calling the system into question. The employee continues to work hard, striving to someday reach that next structural echelon of employment, but the system is never questioned.

With the addition of Althusser's concept of interpellation, I want to move forward to Hall's explanation of the expansion of ideology's production. However, to do justice to the theoretical project of Hall, a few words on Italian Marxist Antonio Gramsci are necessary.

Antonio Gramsci

Antonio Gramsci was an Italian Marxist, revolutionary, journalist (including editor of the socialist journal *L'Ordine Nuovo*), and cofounder of the Italian Communist Party (PCI). In 1911, Gramsci moved from Sardinia in the south of Italy to Turin in Italy's north. This move is more than merely a geographical footnote. The divide within Italy between the industrialized north and the more rural south provided the framework for his insistence on intellectuals emerging from one's class location. Remnants of the "Southern Question" are riddled throughout Gramsci's work. Equally as prescient, during the Red Years of 1918–1920, Gramsci began to struggle with the failure of Left revolutions throughout Western Europe. In the aftermath of the Bolshevik Revolution in Russia, Gramsci grappled with the failure of not only socialist revolutions, but especially how after World War I, the fascist right took power. As Gramsci noted, "the crisis consists precisely in the fact that the old is dying but the new cannot be born; in this interregnum a great variety of morbid symptoms appear."[75]

In late October 1926, an assassination attempt was made on Italian fascist dictator Benito Mussolini's life. Following this event, Mussolini banned and arrested all dissenting and opposition groups. Gramsci was arrested (although he was not involved in the assassination attempt, and had immunity as a member of the Italian Parliament) and in May 1928 he was put on (a show) trial. Gramsci was imprisoned until his release in 1937—he died of complications to a cerebral hemorrhage on April 27, 1937, in the Quisisana clinic in Rome. While in Mussolini's prisons, first at Turi and then in a prison hospital at Formia, and through the constant censor of the prison system (Gramsci was considered a political prisoner due to his involvement with the PCI) Gramsci compiled thirty-three handwritten notebooks—totaling 2,848 pages—containing the sketching of his political thought, posthumously published as the *Prison Notebooks* (*Quaderni del carcere*).

The *Quaderni*, however, were almost lost to history. Gramsci's sister-in-law, with whom he corresponded often while imprisoned, Tatiana Schucht, stole the manuscripts from Quisisana and stored them in a secure vault at Banca Commerciale Italiana before shipping them—via a diplomatic bag, no less—to Moscow. Arguably the most important piece of Western Marxism in the early to mid-twentieth century might have been destroyed. This posthumous fact is crucial for what follows: as Michel Foucault noted, "*C'est un auteur plus souvent cite que reellement connu*," or he is an author cited more than he is actually known.[76] As a journalist, Gramsci published articles during his time (between 1914 and 1918) at *Il Grido del Popolo*, *Avanti!*, *La Città Futura*, and (in 1919–1920) *L'Ordine Nuovo*, but one wonders what he would think of his sketches and blueprints furthering the study of Marxist analysis. What remains, through his complex analysis, is his political legacy.

Gramsci, although working within the broad paradigm of Marxism, "extensively revised, renovated, and sophisticated" that paradigm.[77] Gramsci reformulated the reductive Marxist concept of ideology through his detailed work on hegemony, which Hall describes as an "immense theoretical revolution."[78] Orthodox Marxism posited that humanity—and human will—had little effect on societal change; rather, change came through social and economic processes. The theory of orthodox Marxism is this reductive view that the processes of social change happen independently of human effort. Perhaps more important for where Gramsci takes his own work, at the level of practice, orthodox Marxism was built around inaction. As Professor Bruce Grelle noted, "Marxist theory had largely ceased being understood as a guide for action and had become a scholastic exercise remote from concrete political practice."[79] Gramsci saw that the historical realities failed to match the orthodox Marxist view; instead, he witnessed revolutions in agrarian Russia, but those same revolutionary tendencies failed in Western Europe.

Gramsci developed a more nuanced view of social change as he was drawn to Marx's view of praxis. As Gramsci states, "Marx is the creator of a *Weltanschauung*."[80] Gramsci's view of praxis and political change differed from other Marxists in that he did not think the revolutionary political change that succeeded in Russia would succeed in the West. This assessment was due partly to his concept of consent (one way in which hegemony is established in civil society) and how consent manifests itself in the military metaphor of a "war of maneuver" and a "war of position" (more on this in chapter 6). Further, for a revolution to succeed, the intellectual "common sense" of the people must be critiqued.[81] Gramsci's "common sense" refers to uncritical and conformist ways of understanding the world and one's place within it. Common sense is always "an ambiguous, contradictory and multiform concept, and to refer to common sense as a confirmation of truth is nonsense."[82] Ultimately, this is all an ideological struggle.

Gramsci differs from Marx and Althusser in that he held that ideology could be a positive influence because it binds a group together (although the group itself can have nefarious goals). Ideology as a positive influence molds a faction of people into a group through a solidaristic bond. Gramsci differentiates between this positive conceptualization of ideology and a negative conceptualization of ideology. Gramsci's positive ideology is one that binds historical blocs; the negative ideological conception is one in which ideology returns to ideas that mask exploitation.

Gramsci also breaks from "the austere materialism of Althusser's account of ideology" in that ideology is "the site of disavowal."[83] Disavowal is the liminal space between action and thought; within that space, groups can map the ideological view of another group even if "the social group in question may indeed have its own conception of the world, even if only embryonic. . . But this same group has, for reasons of submission and intellectual freedom and intellectual subordination, adopted a conception which is not its own but is borrowed from another group."[84]

An important distinction in Gramsci's ideology is that it cuts across all levels of society. Ideology, then, as it is analyzed in religious discourses, for example, is not merely for the use of the ruling class. These various religious discourses are used within political struggles. This happens with both the "religion of the intellectuals" as well as the "religion of the people." Within these political struggles is precisely where common sense is dissected.

Gramsci's interest was in how ideology moved from a science of ideas to a specific "system of ideas."[85] Further, Gramsci's nuance is in how he understands the political process. Politics is an ensemble of social relations in which "there really do exist rulers and ruled, leaders and led. The entire science and art of politics are based on this primordial, and (given certain general conditions) irreducible fact."[86] With that, I want to combine the building blocks of ideology, politics, and consent together into the overarching concept of Gramsci's political theory: hegemony. As Terry Eagleton reminds readers, hegemony "*includes* ideology, but is not reducible to it."[87]

Hegemony is the sustaining of political dominance through consent. Therefore, hegemony is always *both* a political development *and* a "politico-practical" concept for Gramsci.[88] Borrowing from Machiavelli, Gramsci asserts that hegemony is similar to a centaur in that the "half-beast and half-man signifies that a ruler needs to use both natures, and that one without the other is not effective."[89] Hegemony, in other words, is a *struggle*, something that must be continuously modified and maintained. Hegemony, like ideology, operates at all levels of a society. The goal of hegemonic powers is essentially to convince the oppressed that what is happening is in their best interests. This process of gaining consent is based on the idea that if one controls an individuals' beliefs or loyalties, one controls the person; this control does not always have to be gained violently, although violence may be used as a last resort. Hegemony is created by the political state to maintain its rule and to produce a domination that is not merely attained through force or the implementation of power, although hegemony is always attained through both consent *and* coercion. On the coercion side is what Gramsci defines as "politico-military" action.[90] The normal conditions of political life will not involve military force or violence. There is balance—an "unstable equilibrium"—between consent and coercion. Gramsci's concept of hegemony explains political struggle in a way that takes other political movements into account. To elaborate, hegemony does not take place in a vacuum. Consent and coercion happen in the context of competition with other worldviews or value systems in life.

Hegemony is spread and further established in a society through the success of intellectuals and the implementation of consent. In Gramsci's schema, an organic intellectual is responsible for organizing and leading from within his or her class structure. Organic intellectuals represent the aspirations of their own class. Organic intellectuals, in Gramsci's impassioned words, *feel* with their class, and "one cannot make politics-history without this passion, without this sentimental connection between intellectuals and people-nation."[91] There is no "elite" detached leadership exercised by an organic intellectual, as one maintains one's role within the class structure. Organic intellectuals ensure that new ideas filter to the masses. These ideas are not forced upon people in a hierarchical, top-down fashion as propaganda; rather, they seep into everyday life as values, language, and culture.

A traditional intellectual, on the other hand, is connected to the previous ruling structure, such as the church, that is now subordinate to the hegemonic ruling class. Teachers and clergy are examples of traditional intellectuals. One final comment about Gramsci's intellectuals: everyone is an intellectual. The difference is that not everyone uses his or her intellectual skill in society, or, as Gramsci noted, "Because it can happen that everyone at some time fries a couple eggs or sews up a tear in a jacket, we do not necessarily say that everyone is a cook or a tailor."[92] Therefore, with that undergirding, I will pause my usage of Gramsci as it is only after we come to understand Hall's work can we offer a Gramscian counterhegemonic proposal.

STUART HALL

Hall is the primary cultural studies conversation partner in this project because he is able to articulate a view of culture and ideology that maps well onto the experience of reintegrating combat veterans. Culture is never a neutral site for reflection; rather, culture is where hegemony is constantly fought. Culture is never static. This battleground, so to speak, offers a space for a counterhegemonic force of resistance—or what Hall often refers to as "interventions." Hall's biography provides important insights into his emphasis on praxis.

Hall was born in Kingston, Jamaica, in 1932, as a "colonized subject" to the crown of Great Britain.[93] In 1951, he left for Oxford, England, on a Rhodes Scholarship. His life, therefore, was an embodiment of colonialism. He lived in the in-between spaces between the two worlds of colonized and colonizer, never fully belonging to either. He left Jamaica as a subject and went to the heart of the "imperial metropole."[94] He never returned to Kingston. Following his undergraduate studies at Oxford, he began a PhD on American-British author Henry James. Hall, however, never completed his doctoral work; instead, he became politically active with organizations such as the Campaign for Nuclear Disarmament (CND).

In addition to his political activism, Hall was instrumental in the development of the New Left in Oxford in 1956. The New Left was an amalgamated Communist organization and a group of Oxford University students. The group was able to occupy a liminal space between orthodox Marxist views of society and the failure of ordinary British parliamentary politics. The New Left was *new* in that they broke away from the imperialist politics they understood to be represented by Great Britain's invasion of the Suez and by the Soviet Union's invasion of Hungary. Hall and the New Left published the *New Left Review*, which followed methodological commitments of Hall's, that is, it relied on Gramsci, Althusser, and other Marxist theorists. Besides these commitments, Hall began to signal where he would eventually take cultural studies and his reliance on popular culture as a site of struggle within politics. In his first editorial in the *New Left Review* he states:

> The purpose of discussing cinema or teen-age culture in NLR [*New Left Review*] is not to show, in some modish way, that we are keeping up with the times. These are directly relevant to the imaginative resistances of people who have to live within capitalism—the growing points of social discontent, the projection

of deeply-felt needs. Our experience of life today is so extraordinarily fragmented. The task of socialism today is to meet people where they *are*, where they are touched, bitten, moved, frustrated, nauseated—to develop discontent and, at the same time, to give the socialist moment some *direct* sense of the times in which we live.[95]

Four years later, Hall coauthored *The Popular Arts*, which further cemented his analysis of the "popular." Also, at this time, he accepted an invitation from Richard Hoggart to go to the University of Birmingham, where Hoggart had started the Centre for Contemporary Cultural Studies (CCCS). The CCCS began as a way to understand the relationship between culture, everyday life, and politics and as a response to the erasure of the British working class in post–World War II Great Britain. In the midst of economic growth and prosperity, important aspects of British cultural life lay dormant. Hall notes that the formation of CCCS was not merely an intellectual project, but rather, it was always a political project of "analyzing postwar advanced capitalist culture."[96]

Hoggart, as the first director of the CCCS, sought to note the cultural changes in Great Britain from his background as a literary critic. Coming from the north of England with a predominately working-class familial background and thrust into the university system, Hoggart experienced, firsthand, the university's bifurcated class lines. Hoggart, and other early CCCS theorists, such as Raymond Williams, took his personal experience and argued from it toward a classless society. Therefore, he extolled the elitism of the British academy and the "high" and "low" culture of Great Britain. The early years of the CCCS sought to dismantle the aristocratic literary tradition of the academy. Perhaps equally as relevant as the class consciousness of early theorists of the CCCS, their methodological commitments proved scandalous to the British academy, and what started strictly as a literary critical methodology crept into a sociological analysis of culture.

During Hall's tenure as the second director of the CCCS from 1968 to 1979 (taking over from Hoggart), a new alignment emerged, more important than the one with sociology, which was the implicit and explicit relationship with Marxism. This is where the CCCS and Hall more specifically connect to this genealogy.[97] Hall helpfully situates the *implicit* reliance on Marxism by the Centre in the context of Cold War Britain and a legitimate fear of working with Marx or Marxist analysis. Hall exegetes sections from Raymond Williams's early texts to show how, for example, when Williams refers to the "system of economic life," or "the system of economic organization" he is referring to Marx's "mode of production."[98] Hall found that the Marxist ideas that influenced the majority of his contemporary intellectual conversations were the same Marx-Engels principle mentioned above and an overall reductive reliance on Marx's base and superstructure concept.

Hall, however, through his reading of Gramsci, was able to usher in a new methodological analysis more amenable to Marxist theory. In particular, Hall was interested in how cultural production manifests a working-class struggle to express itself within the confines of oppression (or what Hall repeatedly refers to as "sites of resistance"). Key insights from Gramsci, especially the practices of hegemony (consent, common sense, and the role of the intellectual) come up constantly in Hall's work. Gramsci, for Hall, signaled a way to utilize Marxist social analysis while discarding the rigid base-superstructure metaphor. Hall's ideology—and in particular, the ideological struggle—is more reminiscent of Gramsci's hegemony.

Returning to culture as a pivot toward chapter 3, it is Hall's view of culture, contra Marx's, which guides this project. Culture is active and represents a formative role within society. Culture is able to do this *through* mass media. Politics reside within culture, and those texts that are inscribed with meaning are media texts. Media, similar to Althusser's ideas, interpellates subjects with a view that subjects are able to freely choose their conditions. Marx would say that it is economics—the base—that determines the media. However, for Hall, the base and superstructure have circuitry between them. Therefore, Hall and other New Left intellectuals argue that culture can also work to influence and determine economics. The superstructure can determine the base, to use Marx's language.

IMPLICATIONS FOR PRACTICAL THEOLOGY AND THIS PROJECT

There are clear advantages to utilizing ideology critique as a conversation partner for military chaplains and practical theologians providing moral injury support. The conceptualization of moral injury support that I am proposing directly confronts the ideological mythologizing of US military service. A dominant ideological view of service members is braided into the media we consume. Ideological images, whether in film or celebrated at sporting events, produce people's understandings of military service (and of war). Hall's theorizing of ideology shows that resistance to hegemonic forces *is* possible, and military chaplains and practical theologians are needed to stand in solidarity with counterhegemonic forces. The significance of utilizing ideology critique (from Hall's cultural studies) as a conversation partner is that culture and media directly impact how veterans' understand their deployment experience. Practical theology is, if nothing else, committed to meeting people where they are. Part of this commitment is a critical engagement with social structures and how these structures impact individuals.

Through this genealogy's theorizing, ideology critique is able to take the chaplain part of the way toward enacting practices of support. That journey of cognition provides ample explanation for ways in which ideology is produced. However, what is missing, and what spiritual care and counseling is uniquely capable of providing, is an understanding of the role of emotion and affect in the power of ideology. Earlier, in the moral injury section of this chapter, I discussed the affects within moral injury symptomology, as proffered by Jinkerson. To recap, core moral injury symptomology includes guilt, shame, spiritual or existential conflict (including "subjective loss of meaning in life"), and a "loss of trust in self, others, and a transcendent or divine entity."[99] At a secondary level, Jinkerson notes the presence of depression, anxiety, anger, a reexperiencing of the moral conflict, self-harm, and social problems.[100]

Now that this genealogy has discussed both moral injury and ideology critique it is clearer how moral injury has an affective ideological element connected to it. The worldview of the military, utilizing Moon's "moral orienting system," is one in which a dominant ideology is produced. Army recruits are inculcated into that ideology through the seven army values: leadership, duty, respect, selfless service, honor, integrity, and personal courage. When, for example, an organizational MIE takes place, a service member is doubly affected because he or she has also failed to live up to the standards of camaraderie. Any treatment intervention that does not include a mechanism for exploring how the ideology of the military impacts the integration of a moral injury is destined to be short-lived. Later I will unpack an alternative framework for the reintegration of veterans with moral injuries within ideological matrices.

Finally, it is not only veterans who are affected by dominant ideologies. I am critiquing how ideology impacts a veteran's return and how this is atomized into the daily interactions between veterans and civilians. On one hand, it is true that "thank you for your service" has become the constitutive statement of American civic religion. For some, then, merely reciting this phrase is enough to fulfill one's civic duty. However, on the other hand, there are civilians desperately looking for ways to support veterans in their communities. Part of the benefit of encouraging veterans to re-author the stories of their MIEs and to share these morally injurious narratives is that civilians will also learn "how American values, logic, and practices regarding suffering and self-sacrifice lie hidden beneath the surface of moral injury."[101]

NOTES

1. Miller-McLemore, *Christian Theology in Practice*, 106.
2. These are two of Richard Osmer's four tasks of practical theology discussed in *Practical Theology: An Introduction* (Grand Rapids, MI: William B. Eerdmans Publishing Company, 2008). I unpack the four tasks below.
3. Gijsbert D. J. Dingemans, "Practical Theology in the Academy: A Contemporary Overview," *The Journal of Religion* 76 (1996): 87.
4. Ibid.
5. *Thick description* is a crucial term, and task, for practical theologians. Anthropologist Clifford Geertz is credited with coining this term in his 1973 text, *The Interpretation of Cultures: Selected Essays* (New York: Basic Books, 1973), 3–30.
6. Andrew Root, "Practical Theology: What Is It and How Does It Work," *Journal of Youth Ministry* 7, no. 2 (Spring 2009): 56.
7. Elaine Graham, Heather Walton, and Frances Ward, *Theological Reflection: Methods* (London: SCM Press, 2005), 3. Emphasis in original.
8. Rebecca S. Chopp, "Practical Theology and Liberation," in *Formation and Reflection: The Promise of Practical Theology*, ed. Lewis S. Mudge and James N. Poling (Minneapolis: Fortress Press, 2009), 135.
9. Karl Marx and Friedrich Engels, *The Marx-Engels Reader*, ed. Robert C. Tucker, 2nd ed. (New York: W.W. Norton & Company, 1978), 145.
10. Although he does not use moral injury terminology, Robert Jay Lifton's 1973 book, *Home from War: Learning from Vietnam Veterans* (New York: Other Press, 2005), does offer similar arguments around moral agency and the effects of guilt and shame. Further, what is helpful for this work is that Lifton critiques how society failed to reintegrate Vietnam veterans.
11. Shay, *Achilles in Vietnam*, 4.
12. Jonathan Shay, "Moral Injury," *Psychoanalytic Psychology* 31, no. 2 (2014): 183.
13. Jonathan Shay, "Moral Injury," *Intertexts* 16, no. 1 (2012): 57.
14. Ibid.
15. S.K. Moore, *Military Chaplains as Agents of Peace: Religious Leader Engagement in Conflict and Post-conflict Environments* (New York: Lexington Books, 2013), 5–6.
16. Litz et al., "Moral Injury," 697.
17. Joseph Wiinikka-Lydon, *Moral Injury and the Promise of Virtue* (New York: Palgrave Macmillan, 2019), 155; "Mapping Moral Injury," 175.
18. Honneth was the director of the (in)famous Institute for Social Research (*Institut für Sozialforschung*), more colloquially known as the Frankfurt School of Critical Theory, from 2001 to 2018.
19. J. M. Bernstein, "Suffering Injustice: Misrecognition as Moral Injury in Critical Theory," *International Journal of Philosophical Studies* 13 (2005): 303–324; *Torture and Dignity: An Essay on Moral Injury* (Chicago: University of Chicago Press, 2015).
20. Wiinikka-Lydon, "Mapping Moral Injury," 180.

21. Andrews spoke of this emerging theme in his research for the Association of Practical Theology, see, https://practicaltheology.org/2015/02/28/ive-researching/.
22. Wiinikka-Lydon, "Mapping Moral Injury," 186.
23. Shay, "Moral Injury," *Intertexts*, 59.
24. Jonathan Shay, "Moral Leadership Prevents Moral Injury," in *War and Moral Injury: A Reader*, ed. Robert Emmet Meagher and Douglas A. Pryer (Eugene, OR: Cascade Books, 2018), 304.
25. Carey et al., "Moral Injury," 1220.
26. D. William Alexander, "Gregory Is My Friend," in *War and Moral Injury: A Reader*, ed. Robert Emmet Meagher and Douglas A. Pryer (Eugene, OR: Cascade Books, 2018), 205n13.
27. William P. Nash and Brett T. Litz, "Moral Injury: A Mechanism for War-Related Psychological Trauma in Military Family Member," *Clinical Child and Family Psychology Review* 16, no. 4 (2013): 368.
28. Ibid.
29. Carey et al., "Moral Injury," 1220.
30. Brock and Lettini, *Soul Repair*, 50.
31. Kent D. Drescher et al., "An Exploration of the Viability and Usefulness of the Construct of Moral Injury in War Veterans," *Traumatology* 17, no. 1 (2001): 10.
32. Jeremy D. Jinkerson, "Defining and Assessing Moral Injury: A Syndrome Perspective," *Traumatology* 22, no. 2 (2016): 125.
33. Ibid., 126.
34. Ibid.
35. Ibid.
36. Duane Larson and Jeff Zust, *Care for the Sorrowing Soul: Healing Moral Injuries from Military Service and Implications for the Rest of Us* (Eugene, OR: Cascade Books, 2017), 20.
37. Jeffrey Pyne and colleagues have recently worked to bring about collaboration between VA chaplains, behavioral health providers, and community clergy in a VA setting. Jeffrey M. Pyne et al., "Mental Health Clinician and Community Clergy Collaboration to Address Moral Injury in Veterans and the Role of the Veterans Affairs Chaplain," *Journal of Health Care Chaplaincy* (August 15, 2018): 1–19.
38. Tom Frame, "Moral Injury and the Influence of Christian Religious Conviction," in *War and Moral Injury: A Reader*, ed. Robert Emmet Meagher and Douglas A. Pryer (Eugene, OR: Cascade Books, 2018), 195.
39. Brett T. Litz et al., *Adaptive Disclosure: A New Treatment for Military Trauma, Loss, and Moral Injury* (New York: The Guilford Press, 2016).
40. Brett T. Litz and Jessica R. Carney, "Employing Loving-Kindness Meditation to Promote Self-and Other-Compassion among War Veterans with Posttraumatic Stress Disorder," *Spirituality in Clinical Practice* (July 12, 2018): 3.
41. Litz et al., "Moral Injury," 701–703.
42. Litz and Carney, "Employing Loving-Kindness," 5.
43. Ibid., 6.
44. Bruce Rogers-Vaughn, "Best Practices in Pastoral Counseling: Is Theology Necessary?" *Journal of Pastoral Theology* 23, no. 1 (2013): 5.

45. Litz and Carney, "Employing Loving-Kindness," 6.
46. J. Irene Harris et al., "The Effectiveness of a Trauma-focused Spiritually Integrated Intervention for Veterans Exposed to Trauma," *Journal of Clinical Psychology* 67, no. 4 (2011): 425.
47. J. Irene Harris et al., "Moral Injury and Psycho-Spiritual Development: Considering the Developmental Context," *Spirituality in Clinical Practice* 2 (January 1, 2015): 262.
48. Larry Kent Graham, *Moral Injury: Restoring Wounded Souls* (Nashville: Abingdon Press, 2017), 39.
49. Ibid., 109.
50. Ibid., 11.
51. Ibid., 39.
52. Carrie Doehring, "Military Moral Injury: An Evidence-Based and Intercultural Approach to Spiritual Care," *Pastoral Psychology* 68, no. 1 (February 2019): 15–30; Joshua T. Morris, "'Thank You for Your Service:' Mapping Counter-Memories as a Form of Spiritual Care Support for Moral Injury," *Journal of Pastoral Theology* 28, no. 1 (2018): 34–44.
53. Zachary Moon, *Coming Home: Ministry That Matters with Veterans and Military Families* (St. Louis: Chalice Press, 2015).
54. Moon, *Warriors Between Worlds*, 3.
55. Ibid., 4.
56. Graham, *Moral Injury*, 133.
57. Ibid.
58. Gayatri Chakravorty Spivak, "Can the Subaltern Speak," in *Colonial Discourse and Post-Colonial Theory: A Reader*, ed. Patrick Williams and Laura Chrisman (New York: Columbia University Press, 1994), 68.
59. Marx and Engels, *The Marx-Engels Reader*, 150. Emphasis in original.
60. Ibid., 157.
61. Ibid., 154.
62. Stuart Hall, "Culture, the Media, and the 'Ideological Effect,'" in *Mass Communication and Society*, ed. James Curran, Michael Gurevitch, and Janet Woollacott (London: Sage Publications, 1977), 318.
63. Marx and Engels, *The Marx-Engels Reader*, 154.
64. Ibid., 172–173.
65. Hall, "Ideology and Ideological Struggle," 127.
66. Louis Althusser, "Ideology and Ideological State Apparatuses (Notes toward an Investigation)," in *Lenin and Philosophy and Other Essays*, trans. Ben Brewster (New York: Monthly Review Press, 1971). This essay is one of the most important pieces in cultural studies.
67. Henry Krips, "Ideology and its Pleasures: Althusser, Žižek, and Pfaller," *Continental Thought & Theory: A Journal of Intellectual Freedom* 2, no. 1 (June 2018): 336.
68. Ibid. Emphasis in original.
69. Althusser, "Ideological State Apparatuses," 166.
70. Hall, "Ideology and Ideological Struggle," 131. Emphasis in original.

71. Althusser, "Ideological State Apparatuses," 142–143.
72. Ibid., 156.
73. Ibid., 174.
74. Ibid., 162.
75. Antonio Gramsci, *Selections from the Prison Notebooks*, ed. and trans. Quintin Hoare and Geoffrey Nowell Smith (New York: International Publishers, 2014), 276. The failure of revolution to take hold in western Europe is a catalyst for Gramsci's *Prison Notebooks*. Similarly, the Frankfurt School of Critical Theory emerged in the aftermath of World War I, through the financial backing of Hermann Weil to initiate a quasi-Marxist think tank dedicated to explaining the failure of revolution in Germany.
76. Joseph A. Buttigieg, "Preface," in Antonio Gramsci, *Prison Notebooks*, ed. and trans. Joseph A. Buttigieg, vol. I (New York: Columbia University Press, 1992), xix.
77. Stuart Hall, "Gramsci's Relevance for the Study of Race and Ethnicity," in *Stuart Hall: Critical Dialogues in Cultural Studies*, ed. David Morley and Kuan-Hsing Chen (New York: Routledge, 1996), 411.
78. Hall, "'Ideological Effect,'" 334.
79. Bruce Grelle, *Antonio Gramsci and the Question of Religion: Ideology, Ethics, and Hegemony* (New York: Routledge, 2017), 15.
80. Gramsci, *Selections*, 381.
81. A contemporary example of common sense would include the accepted vernacular of the differentiation between the wealthy considered the "1 percent" while the rest of society make up the "99 percent." Prior to the Occupy Wall Street movement and moment, one would have not have heard these descriptors. However, what the Gramscian intellectuals of the Occupy movement, such as David Graeber, achieved was imbuing the culture with new ideas and vocabulary concerning how the world functions. Those designations became a mainstay of Vermont senator Bernie Sanders's 2016 and 2020 presidential campaigns.
82. Gramsci, *Selections*, 423.
83. Krips, "Ideology and its Pleasures," 338.
84. Gramsci, *Selections*, 327.
85. Ibid.
86. Ibid., 144.
87. Eagleton, *Ideology*, 112.
88. Gramsci, *Selections*, 333.
89. Niccolò Machiavelli, *The Prince*, ed. Quentin Skinner and Russell Price (New York: Cambridge University Press, 1988), 61.
90. Gramsci, *Selections*, 107, 183.
91. Ibid., 418.
92. Ibid., 9.
93. Stuart Hall, *Familiar Stranger: A Life Between Two Islands*, ed. Bill Schwarz (Durham, NC: Duke University Press, 2017), 3.
94. Ibid., 10.

95. Stuart Hall, "Introducing NLR," *New Left Review* 1, no. 1 (January–February 1960): 1.
96. Stuart Hall, "The Formation of Cultural Studies," in *Cultural Studies 1983: A Theoretical History*, ed. Jennifer Daryl Slack and Lawrence Grossberg (Durham, NC: Duke University Press, 2016), 7.
97. In 1979, Hall left the CCCS to become Professor of Sociology at the Open University, where he stayed until his retirement in 1997. The Open University, established in 1969, sought to provide high-quality education that is flexible and provided at a distance. The Open University recorded lectures on cassette tapes and video-recorded them on VHS tapes in conjunction with the BBC.
98. Hall, "The Formation of Cultural Studies," 21.
99. Jinkerson, "Defining and Assessing Moral Injury," 126.
100. Ibid.
101. Kelly Denton-Borhaug, "Like Acid Seeping into Your Soul: Religio-Cultural Violence in Moral Injury," in *Exploring Moral Injury in Sacred Texts*, ed. Joseph McDonald (London: Jessica Kingsley Publishers, 2017), 119.

Chapter 3

Hermeneutical Circles and Liberative Praxis

The methodological approaches that guide this book are twofold: first, a practical theology hermeneutical method of liberative praxis; second, the qualitative methodology of case study. To articulate a liberative praxis method, I will trace how Juan Luis Segundo's method of a liberative hermeneutical circle is a complementary way of employing hermeneutics and, in particular, when combined with pastoral and practical theologian Emmanuel Lartey's liberative praxis. Building on Segundo's method, I argue that the use of case study as a qualitative methodology is tenable within this strand of practical and pastoral theology and liberative praxis. With that, I want to now trace and develop my practical theological method.[1]

LIBERATIVE PRAXIS

The four tasks of practical theology assist in identifying frustrations with first-generation moral injury research. Although American practical theologian Richard Osmer is known for his systematizing of the theological tasks, he states, "I make no claim to originality in my description of these tasks. While the terms may differ, something like each of them is taught in clinical pastoral education, doctor of ministry courses, and courses on preaching, pastoral care, administration, Christian education, and evangelism in schools of theology."[2] Osmer designates the four tasks as follows: the "descriptive-empirical," the "interpretive," the "normative," and the "pragmatic."[3] In the first task, the practical theologian is seeking information to discern what is going on. In the interpretative task, in conjunction with interdisciplinary conversation partners, the practical theologian explores and explains what is going on. So, for example, in this book, I am particularly exploring *why* society has failed to fully reintegrate veterans.

The remaining two tasks are what provide practical theology with its impetus—and what I argue are upheld in this project. Moving into the normative task, the researcher's interpretive focus turns to the episodes, situations, and contexts in dialogue with theological texts and begins to develop—and co-construct—practices in response. Osmer's hermeneutical circle is located in incidents (what he refers to as "episodes") within patterns and relationships ("situations") and within an understanding of the overall systemic implications of the particular event ("context"). These practices are normative as they are generated from the context in question. Further, within this task, the practical theologian names personal norms, choices, and preferences that he or she privileges in providing care. With the pragmatic task, the question is how to move forward faithfully and in connection to the context that precipitated the inquiry.

As I detailed in chapter 2, much of moral injury research addresses Osmer's first two tasks of practical theology as clinicians, theologians, ethicists, and others situate and differentiate moral injury from other combat phenomena (most notably, PTSD). The hermeneutical task is quite useful, and research from clinical psychologists is helping shape our collective understanding of why MIEs impact certain individuals. However, I leave moral injury research wanting to know *more* about how I, as a chaplain, can pragmatically provide care to veterans suffering from such injuries. There has been work from the psychotherapeutic perspective on treating MIEs—most notably AD and AD-E—but this ultimately does not fully assist chaplains (who do not have psychotherapeutic training or additional behavioral health training) in providing competent spiritual care. BSS does, however, provide this sort of training.

Further still, from a liberation theology perspective, unless the system that undergirds these phenomena is critiqued, there is little reason to believe that MIEs will decrease, as ambiguous military conflicts will continue. Segundo speaks to the consequences of an unwillingness to critique systems: "Theology will become and remain the unwitting spokesman of the experiences and ideas of the ruling factions and classes."[4] To prevent becoming an "unwitting spokesman" of the ruling classes, spiritual caregivers need to adopt a liberative praxis method of practical theology. Liberative praxis provides caregivers a framework to provide competent and clinically astute care to individuals, and it provides this care within a broader understanding and critique of how systems of oppression subjugate individuals. To make this argument more completely, I want to offer first a description of the development of practical theology toward a liberative praxis and, in particular, emphasize the hermeneutical circle employed by Juan Luis Segundo and Emmanuel Lartey. My goal is to unpack what is going on in my practical theological method: First, I add Segundo's hermeneutical circle outlined in *Liberation of Theology* as a preferred usage of the hermeneutical circle.

Second, I expand upon pastoral and practical theologian Emmanuel Lartey's liberative praxis method as a way to enter the narratives of this project.

The Liberative Praxis and Hermeneutical Circles of Juan Luis Segundo

Juan Luis Segundo's, *Liberation of Theology* is directly influenced by (as well as influenced) Latin American liberation theology. Further, from these influences come the spiritual care and counseling practices that offer solidarity to veterans. For this genealogy, I am going to mark the beginning of Latin American liberation theology as a discipline with the 1968 Second Conference of Latin American Bishops at Medellín, Colombia (CELAM), which itself emerged from the trajectory of the Cuban Revolution and Vatican II. Concurrently, Black liberation theology was emerging in the United States. The political and social consciousness generated from the civil rights movement and the assassinations of Malcolm X in 1965 and Martin Luther King Jr. in 1968 provided significant momentum in solidifying a constructive Black liberation theology. In other words, the stage was already set to develop a theology that speaks to—and from—communities of the oppressed. For example, in the academy, James Cone's 1969 *Black Theology and Black Power* brought to the fore the emerging theme of Black power, which was already emerging as a motif in the Black community with the activism of such individuals as Stokely Carmichael. What emerged in the academy first gained traction in the community. In 1970, Cone published *A Black Theology of Liberation*, further cementing the hermeneutical and systematic treatment of liberation in theology.

For Latin American liberation theology, broadly speaking, a primary goal of ministry and theological reflection is solidarity with the oppressed, a recognition of God's preferential option for the poor and marginalized, and a transformation of oppressive systems. To meet that goal, Latin American liberation theologians, including Segundo, employ a Marxist praxis. As I emphasized in chapter 2, to truly understand praxis from a Marxist perspective, one must understand that it is never merely a reflection on practice; it must involve change, or "revolution." This happens through a dialectic of practice, reflection on practice, and the synthesis of praxis. The material interests of the ruling class drive society, and therefore, these interests determine how social relations function. To be emancipated from this situation, the goal is not to engage abstractly with the ruling class but to struggle for change through an elevated self-consciousness. In other words, the goal is not *orthodoxy* of right belief; rather, the goal is an *orthopraxis* of changed behavior. Therefore, as Marx himself stated, "The philosophers have only interpreted the world in different ways; the point is to change it."[5] Remembering this purpose of Marxist analysis can help us unpack the meaning of Segundo's hermeneutical circle.

Segundo begins *Liberation of Theology* by arguing that some theological institutions (for him, the Roman Catholic Church) merely appropriate liberation theology terminology, such as *liberation*; these institutions may appropriate *liberation* as content, but they fail when one interrogates their methods. He noted, "it is the methodology that ensures that the existing system will continue to look like an oppressor on the horizon of theology itself; that offers the best hope for the future of theology."[6]

Segundo's hermeneutical circle is "the continuing change in our interpretation of the Bible which is dictated by the continuing changes in our present-day reality, both individual and societal."[7] It is a circle of interpretation as each new piece of data (a part) forces us to reinterpret the tradition (the whole). Important for Segundo is that the methodology he envisions is a revised praxis method of correlation, which is based on Segundo's concept that a priori to the hermeneutical circle is the understanding that sacred texts and traditions can and must be changed through fresh interpretations based in problems and life experiences. Any theology that refuses to change in this hermeneutical way is "conservative" in the sense that "it lacks any *here-and-now* criteria for judging our real situation."[8]

Segundo's hermeneutical circle has four elements. The first is that there is a hermeneutic of suspicion in the way we understand reality. This hermeneutic of suspicion is built upon two preconditions. The first precondition is that the questions rising from lived experience are existential and speak to actual lived experience because those are the questions that one can bring to theology. The second precondition is embedded within a revised praxis method of correlation. Segundo is adamant that if theology responds to lived experience without changing preexisting interpretations, then "that immediately terminates the hermeneutic circle."[9] New questions emerging from lived experience demand that theology take them seriously and not rely on former ways of understanding. It is necessary to go back to our religious resources and reinterpret, mold, and fashion a response.

The second element is that suspicions are interrogated through our "ideological superstructure" and theology.[10] The ideological superstructures are in place to make sense of marginalization, oppression, and subjugation. Segundo's second element gets close to resembling a Marxist ideology critique: the place of ideological superstructures is to "unmask the reality of oppression."[11] Third, the ideological suspicion morphs into an exegetical suspicion of how the sacred texts have been interpreted. What has been missing in interpretation? How can the sacred texts morph to encompass these new concerns? Finally, one emerges from exegetical suspicion with a renewed interpretation of both the experience and the sacred text.

For a method that Segundo states needs to have common sense and not reside solely in the academy, his elements are arduous, so I want to offer one of his own examples as a means of unpacking his steps. Segundo offers four examples, but it is his fourth example, James Cone's *A Black Theology of Liberation*, that remains the example par excellence of his fourfold hermeneutical circle. First, Cone's text starts with his own personal experience. One cannot block out one's history. Cone's partiality is evident in his explication of God's ontological solidarity with the Black community. God is with the oppressed, as God *is* oppressed. Cone is responsible to the Black community; he is not concerned with white theology or its interpretation.

For Segundo's second stage, suspicions are interrogated via our "ideological superstructure" and theology; thus, Cone interrogates the racism and white supremacy of the United States. Even more specifically, Cone critiques the "color-blind" theology of white supremacy.[12] Since Cone started from his own partiality, he has no need for a colorless God or a theological system that espouses a colorless God; a colorless God is merely the oppressor failing to take the cause of racism into account. Segundo's third stage, a new exegetical suspicion, works to unmask the racism in theology and to offer new theological accounts. Cone does this with a continued reliance on the Exodus narrative in the Hebrew Scriptures. Cone arrives at the fourth stage, a new interpretation of the tradition or sacred texts, by stating that God is with the oppressed through the event of the crucifixion and resurrection of Jesus Christ. Christ is not "confined to the first century, and thus our talk of him in the past is important only insofar as it leads us to an *encounter* with him *now*. . . . I want to know what God's revelation means right now as the black community participates in the struggle for liberation."[13]

If I could condense Segundo's hermeneutical circle down to one word, I would decidedly define it as *orthopraxis*—especially over and against orthodoxy. Segundo's hermeneutical circle comes to some semblance of truth only through practice and commitment. On this point, Segundo states, "The truth is truth only when it serves as the basis for truly human attitudes."[14] Borrowing from Cone, Segundo suggests that orthodoxy is in service to orthopraxis. Truth emerges when it serves human needs. It is only through orthopraxis that theology can retain its revolutionary call. Otherwise, "Theology will become and remain the unwitting spokesman of the experiences and ideas of the ruling factions and classes."[15]

A liberative methodology begins with commitment and then moves to theology. Segundo is reliant on what he understands as Jesus's own methodology, which was focused primarily on the human condition, and on committing himself to that task. Segundo notes that Jesus's primary concern was bringing "remedy to some sort of human suffering."[16] It was the Pharisees, Segundo asserts, who began with theology.

I want to discuss Segundo's use of ideology, especially as it directly relates to my central concerns. How does Segundo's liberation theology relate to ideology critique? First, and most important, Segundo does not follow the negative conception of ideology as the distorted ideas that mask the real conditions of a system and that work to benefit those in power. Segundo, in the midst of a text that is properly situated as political, fails to maintain and uphold the political dimension of ideology. Rather, he attempts a "re-ideologization" in which ideology is "the system of goals and means that serves as the necessary backdrop for any human option or line of action," and "a logical system of interconnected values."[17] There is a sense in Segundo's work that ideology is efficacious; it provides the undergirding for action. This means, clearly, that ideology can be constituted as a positive or as a negative force. It is at this point that Segundo introduces "faith." Faith is the "total process to which man submits, a process of learning in and through ideologies how to create the ideologies needed to handle new and unforeseen situations in history."[18] For Segundo, faith and ideology need each other; there is a unifying dimension in their relationship, and they are linked to bring about liberation for the marginalized and oppressed. Ideology, for Segundo, is the way in which oppressed communities move toward actualizing their emancipation.

Segundo, unlike the ideology critique theorists detailed in chapter 2, is not arguing to unmask ideology, but rather to find strategic ways to maximize one's faith to create new ideologies that move people toward emancipation. Segundo seeks to use ideology *better*. Faith represents the "permanent and the unique" while ideologies represent "different historical circumstances."[19] Said differently, faith aligns with an ideology that is better positioned to operationalize the tasks of emancipating marginalized communities. According to Segundo:

> Faith then is not a universal and atemporal pithy body of content summing up divine revelation once it has been divested of ideologies. On the contrary, it is maturity by way of ideologies; the possibility of fully and conscientiously carrying out the ideological task on which the real-life liberation of human beings depends.[20]

One of the primary cognate interlocutors from chapter 2, Hall, is a perfect supplementary conversation partner for Segundo, and this directly relates to Segundo's method. Hall's "hypothetical" reading of ideological texts (which I explore in much greater detail in chapter 5) provides an analysis of ideology critique that can offer liberation theology the critical impetus necessary to have a more holistic critical theory of systems of oppression. Without acknowledging ideology's power, especially in the hands of hegemonic power structures, theology does a disservice to the communities it claims to represent. A political awareness of ideology is necessary to maintain a realistic vision of how hegemony is continuing to maintain power and oppress those on the margins. Segundo's refusal to hold a view of ideology that recognizes its production by hegemonic powers becomes the very barrier to emancipation for marginalized communities. So, while Segundo offers a hermeneutical circle that is committed to the margins and the oppressed, it fails at recognizing the actual dominating forces of ideology. Some of these concerns are addressed in Lartey's intercultural model of liberative praxis, to which I now turn.

Emmanuel Lartey's Liberative Praxis

As it applies to this project, I understand practical theology as an enterprise that privileges the lived experiences of veterans with moral injuries who are reintegrating into society. As the last section on the hermeneutical circle showed, my understanding of a preferred spiritual care paradigm is grounded within practical and pastoral theology's hermeneutical approach. Ultimately, my liberative praxis seeks to privilege those on the margins of systems. My interpretive lens within practical and pastoral theology is what allows me to conceptualize the implicit complications of a US ideology of military service that cannot create space for reintegration. One conversation partner who provides a spiritual care understanding of praxis is Emmanuel Y. Lartey. Lartey's work expands the intercultural perspective while adapting the traditional hermeneutical circle of action–reflection–action or see–judge–act, and he offers a five-stage liberative method for practical theology. His method moves through concrete experience, situational analysis, faith perspectives, interrogation of the situation, and response; this process can be entered at any phase.[21]

Lartey starts with concrete experience. An intercultural approach to a liberative praxis must be "inductive, collective, and inclusive."[22] Caregivers are compelled to engage with a local community, thereby avoiding overgeneralizations about the community. In the context of moral injury, this once again privileges the story of the veteran. The stage of concrete experience allows for real encounters with actual people. From that starting point, Lartey advocates a next step of situational analysis. In this step, the workings are those of a revised praxis method of correlation, as they invite a reciprocal critique of the situation from other disciplines. Lartey acknowledges the limits of this step, and therefore gives a perspectival view, but he insists that a "collective seeing," a comparing of visions, is critical.[23] His third step, a theological analysis of various faith perspectives, allows for religious traditions to have a say, both in questioning and in responding to concrete experiences and concerns. Faith traditions are challenged and critiqued in Lartey's method. In the fourth phase, interrogation of the situation, faith traditions are pushed to answer the questions that arise from the concrete experiences. In other words, the pragmatic and normative tasks of Osmer's practical theology are highlighted. Finally, the responses are considered, and new practices are enacted.

This book utilizes practical theology to understand how veterans in different locations and under divergent societal circumstances make meaning from their moral injuries and, further, how meaning-making is shaped by dominant ideologies. This starting point relies on the hermeneutical task of studying what Miller-McLemore calls the "living human web." The concept of humans existing within webs compels me to develop a thick description of my research participants, which I do in chapter 4. To properly reintegrate and care for returning Reserve component veterans, this thick description must reckon with broader religious, social, and cultural implications. With that said, to meet this commitment, in this project I employ case study to more fully understand the experience of a subset of returning veterans.

CASE STUDY

The method of qualitative inquiry for this study is case study, and employing it as my qualitative methodology is useful as it provides the opportunity to reflect deeply and critically on the essence of reintegrating with an MIE. Case study provides the rich particularity of the lived experience of the veterans' stories while also highlighting the complexity of the interdependency of various influences. Through this in-depth level of analysis, I was able to learn how a dominant ideology functioned for my participants' reintegration.

Case study is a bounded study of an object (a phenomena or phenomenon), rather than a process. The case is bounded to signify that the bounded system has a "boundary and working parts. . . [It] is likely to be purposive, even having a 'self.'"[24] The case, then, is used to study something else, something broader. Researchers refer to this as an "instrumental case study." A simplified way to differentiate between the types of case study is the focus: an intrinsic case study is centered on the case, while in an instrumental case study the "issues" determine the focus, thus this project is an instrumental case study.[25] Case study is amenable to this project and has been a staple within pastoral theology.

PASTORAL THEOLOGY'S OPERATIONALIZATION OF CASE STUDY

Anton Boisen is credited as the founder of CPE, an educational program in which theological students or ministers are placed in a professionalized setting and receive training through supervision and small group reflections. Most of the learning, however, takes place through *doing* pastoral care in the various contexts of the program. Boisen's passion for this model came as a result of his own dissatisfaction as a psychiatric patient. Throughout his vulnerably written, memoir-like text (arguably his own case study), *The Exploration of the Inner World: A Study of Mental Disorder and Religious Experience*, Boisen discusses what it was like to be hospitalized and his experience of what he describes as "disturbances."[26] Much of what he saw around him was the result of spiritual difficulties, and the medical system at that time was not able to address these concerns clinically.

Boisen's observations are critical. His thesis centers on a patient's "insanity" coming from one of two sources: either "organic trouble" (what Boisen understood as a brain defect) or a "disorganization" of the patient's world.[27] These existential disorganizations were not treated properly because they were not treated spiritually. At this time, hospital chaplains—as we understand the role today—were not common. Boisen describes his second disturbance as occurring after listening to a civilian pastor lead chapel services. Their ministry was limited to the hour or so that they provided a service; they were not permitted to visit the patients.

In 1922, Boisen began to work alongside Harvard University physician and ethicist Dr. Richard Cabot, who introduced him to the case study method. The case study methodology is arguably Dr. Cabot's greatest influence on Boisen. It was Cabot's published medical work (not necessarily his social ethical work), in which he and other medical students worked through case study material to emphasize differential diagnoses for different patients, that impressed Boisen. He understood that it was through the study of a living human document that one could make a diagnosis.

With Dr. Cabot, Boisen put together an experimental plan for a chaplain training program. The students in this program worked *in* the hospital and had access to the patient's medical records. Access to medical records, what he referred to as "first-hand sources," was non-negotiable for Boisen; to do this work well, a student needed to know the full case history of a care receiver.[28] The medical records represented the case study for review. Boisen states, "I wanted them to learn to read human documents as well as books, particularly those revealing documents which are opened up at the inner day of judgment."[29] Case study and the subsequently added verbatims are now staples of modern CPE curricula and experiences.

What I think is often missed with Boisen is that he was interested in sociocultural situatedness. He states that the study of the "living human document" is always in conjunction with "actual social locations in all their complexity," and therefore it is these two together that provide the foundation of Boisen's method.[30] For him, human experience needs to be read in the same manner, and with the same rigor, as classical biblical texts. Building on this, "We have sought to determine the origin and meaning of these beliefs, their function in the individual's life, and their implications for a general system of values."[31] To take this seriously, Boisen elaborated on the case study methodology.

Pastoral theologian Charles V. Gerkin took Boisen's living human document idea and added the hermeneutical import that the field of pastoral theology now associates with the phrase. Gerkin understood that pastoral counselors "are more than anything else, listeners to, and interpreters of stories."[32] Stories matter, and language is how we construct meaning. Not only the elements contained within a story, but also the interpretation and re-speaking of a story by a pastoral counselor hold the possibility of providing healing. Gerkin encouraged pastoral counselors to approach people's stories as a "stranger." Gerkin stated, "To listen to stories with an effort to understand means to listen first as a stranger who does not yet fully know the language, the nuanced meanings of the other as his or her story is being told."[33] The capacity to understand another person is, for Gerkin, a hermeneutical ability. An important element of Gerkin's thought is that this hermeneutical process is an intersubjective task. The mutuality that is present between a caregiver and a care receiver is the result of the melding of two living human documents who have interpreted one another and made meaning together.

Both Boisen and Gerkin provide invaluable additions to the field of pastoral theology and the method of case study. They provide a voice advocating for the marginalized and oppressed. However, although Boisen in particular addresses sociocultural concerns, the concept of a living human document remains too narrow and fails to fully conceptualize social and political implications for both care providers and care receivers. Pastoral theologian Glen Asquith remarks that Boisen's legacy "may well be the insight that one must take the time and discipline to ask the right questions in order to obtain a more complete theological perspective on human situations."[34] Boisen was able to achieve a milestone: access to patients and their records. Yet this leads to empiricism and an inherent hierarchy between a theological student and a patient. It is within this trajectory that Miller-McLemore's living human web, which I addressed in chapter 1, becomes helpful.

To recap, Miller-McLemore's concept of the living human web takes seriously the concerns of marginalized communities, and it moves away from the narrowly defined "living human document" to a broader, more critical focus on the contextual aspects of care. Her conceptual contribution to the practical theology field focuses on the embeddedness of persons in various public webs of meaning. Importantly, Miller-McLemore is not advocating the complete erasure of the living human document as an image by which to understand care; rather, she has a broader conceptualization of care that takes the holistic approach of putting many multifaceted systemic realities into play. "Ultimately," she says, "I suggest that 'the living human document within the web' is the metaphor that best captures the subject matter of both CPE and pastoral theology."[35]

IMPLICATIONS FOR PRACTICAL THEOLOGY AND THIS PROJECT

Methodologically, this project is committed to a practical theological liberative praxis. Lartey's praxis, in particular, is essential for providing effective spiritual care that takes the web of interdependence seriously. To take these webs and social systems seriously, I utilize case study. Looking at social systems, I am concerned with how dominant ideologies exacerbate and complicate reintegration with moral injuries; this is the ideological critique I am engaging. Through this excavation, we are firmly placed to hear the stories of my participants.

NOTES

1. Please see the Appendix for an analysis of my research design.
2. Osmer, *Practical Theology*, 4.
3. Ibid.
4. Segundo, *Liberation of Theology*, 39.
5. Marx and Engels, *The Marx-Engels Reader*, 145.
6. Segundo, *Liberation of Theology*, 40.
7. Ibid., 8.
8. Ibid., 9.
9. Ibid.
10. Ibid.
11. Ibid., 27.
12. Ibid., 28.
13. James H. Cone, *A Black Theology of Liberation* (New York: Orbis Books, 2010), 31. Emphasis in the original.
14. Segundo, *Liberation of Theology*, 32.
15. Ibid., 39.

16. Ibid., 79.
17. Ibid., 116, 102, 105.
18. Ibid., 120.
19. Ibid., 116.
20. Ibid., 122.
21. Emmanuel Y. Lartey, *In Living Color: An Intercultural Approach to Pastoral Care and Counseling*, 2nd ed. (London: Jessica Kingsley Publishers, 2003), 131–133.
22. Ibid., 125.
23. Ibid., 132–133.
24. Robert E. Stake, *The Art of Case Study Research* (Thousand Oaks, CA: Sage Publications, 1995), 2.
25. Ibid., 16.
26. Anton T. Boisen, *The Exploration of the Inner World: A Study of Mental Disorder and Religious Experience* (Philadelphia: University of Pennsylvania Press, 1936).
27. Ibid., 10–11.
28. Ibid., 10.
29. Ibid.
30. Ibid., 185.
31. Ibid., 306.
32. Charles V. Gerkin, *The Living Human Document: Re-Visioning Pastoral Counseling in a Hermeneutical Mode* (Nashville: Abingdon Press, 1984), 26.
33. Ibid., 27.
34. Glenn H. Asquith, "The Case Study Method of Anton T. Boisen," *Journal of Pastoral Care* 34, no. 2 (June 1980): 94.
35. Miller-McLemore, *Christian Theology in Practice*, 51.

Chapter 4

The Reification of the Veteran
Kaleidoscopic Lived Experiences

This chapter unpacks the thick description of the reintegration experience of Phillip Campbell, Lisa Fisher, Angela Gallagher, and Andrew Lloyd, all of whom are veterans of the wars in Iraq or Afghanistan. Returning to the four tasks of practical theologian Richard Osmer, much of this chapter engages in his first two tasks: the "descriptive-empirical" and the "interpretive."[1] Following this chapter, chapters 5 and 6 will take this interpretative work and move toward Osmer's "normative" and "pragmatic" tasks, although some foreshadowing to these tasks is present here as well.

I now proceed into describing the unique situatedness and braided social locations of my participants that offered the context within which my study unfolded. Following closely from there, I will detail three prominent themes of alienation that I uncovered via my coding: belonging, divided identities, and betrayal. Finally, this chapter will detail some of my participants' religious resources; this transitions the project into chapter 5 and its focus on a revised praxis method of correlation, using my theological and cognate conversation partners. With this structural road map in place, I now introduce my participants.

U.S. ARMY RESERVE SPECIALIST PHILLIP CAMPBELL

Specialist Phillip Campbell is in his late forties. Phillip is a straight, white, and cisgender male. He is the father of two adult children (ages 22 and 24), and a grandfather to a two-year-old granddaughter. Phillip and his wife Vicki have been married for eleven years, including a four-month separation after his deployment to Afghanistan in early 2013. Phillip led an inconspicuous life before enlisting in the military. He described it as a time of "dead-end jobs" and of ultimately "going nowhere with my life." Phillip was able to work some private security jobs in his early thirties, which gave him a sense of doing something that mattered. He would iron his uniform, polish his boots, and clean his pistol with a level of dedication that would eventually help him excel in the army. He felt a need to "serve his country" in the post-9/11 Global War on Terror landscape, which led him to start asking deeper questions about perhaps enlisting in the military.

Pragmatically, apart from any patriotic pride, Phillip and Vicki wanted to own a home, and they knew that without a VA Home Loan, they would probably never be able to afford one. This was in 2010, on the heels of the 2008 US economic collapse. Phillip knew in his head that the army was the right place for him; he was tired of trying to balance his checkbook and secure a decent mortgage. Phillip's story is increasingly common for some service members in the Reserve component: there is a healthy mixture of pragmatism and patriotism that makes serving part-time in the military an honorable option.

When it came time to decide what he was going to do in the military, he was just "excited, stoked even" to serve, and therefore he wanted to serve in various jobs, or "Military Occupational Specialties" (MOS), including the military police, chaplain's assistant, or truck driver. Phillip drove his old pickup truck home from the recruiter's office and reflected on growing up working on trucks with his dad. He returned to the recruiter the next day and decided on becoming a truck driver, a "Motor Transport Operator," which is more commonly referred to by its MOS designator as an "88M" or "88 Mike."

Despite the details I will unpack momentarily, Phillip's military service remains a proud moment of his life. To this day, he often walks down into his basement, where his army kit is located. He sits and gazes at his uniform, making sure there is no dust on any item. His uniforms lie crisply on a hanger, displayed with reverence. In his basement are other mementos: ID tags (colloquially known as "dog tags"), unit crests, photos from combat, and a DVD that his unit compiled of their tour. He still watches it yearly. It has become an early summer tradition. Phillip started smoking cigars in Afghanistan, and he continues that habit at home; every time he has a "cigar night," he wears his black mesh cap with a Velcro patch of his platoon's nickname: The Wolf Pack.

U.S. ARMY CORPORAL LISA FISHER

Corporal Lisa Fisher is in her mid-twenties and enlisted in the U.S. Army after a tumultuous home life. Yes, her enlistment mirrored the military brochure reasons—she wanted health benefits, tuition assistance for college, and a way out. The military provided stability. Lisa experienced homelessness after being kicked out of her home at sixteen. Her late adolescence entailed couch-hopping and sneaking into high school bathrooms to find the time to bathe. A female recruiter came to her high school and explained her potential options. It was not that the military was never an option for her; she held service members in some degree of esteem and even wrote letters to some who were deployed. It was the fact that this recruiter was a woman—and of small stature—that Lisa *felt* she too could enlist.

Lisa deployed to Afghanistan with a combat support battalion. Beyond her normal duties, she held additional duties at the Combat Support Hospital (CSH). The CSH provided an opportunity to quietly serve and contribute differently to the war effort. During a Sexual Harassment Assault Response Prevention (SHARP) training she listened as senior ranking officers and noncommissioned officers (NCOs) discussed a recent case in which a female service member video recorded a senior officer sexually harassing her. Lisa thought, for once, the men in the room would understand what women face. However, the discussion quickly turned to victim blaming: although recorded, what did the woman *do* to the man that made him advance on *her*? The lessons from this session would remain cemented in her psyche: if you report, the army will not believe you. Lisa had her own experiences of sexual harassment and assault. Whether it was the constant groping while washing her hands at the dining facility (DFAC) or while standing at Parade Rest, itself a submissive posture, when her First Sergeant said he would not recommend her subsequent assignment unless she slept with him. "Well," she thought, "he can fuck all the way off, at least I am not homeless."

U.S. ARMY RESERVE PRIVATE FIRST CLASS ANGELA GALLAGHER

Private First Class (PFC) Gallagher enlisted in the Army after high school. She enlisted with a sense of patriotism to her country. Her father and grandfathers served as well. She also grew up in San Diego, itself a highly concentrated military community. She deployed to Afghanistan shortly after completing her Advanced Individual Training (AIT) as a Human Resources Specialist (42A). In hindsight she is able to comment that the deployment was "pretty boring, actually." She did her job, traveled to support the battalion's mission, but she was surprised that the mission did not have the images conjured up by movies—"when are we going to kick in doors?" However, it is only in hindsight that she realizes this. That season of her life, and immediately following her deployment, contained some of the most traumatic and chaotic situations she has faced.

She left for Afghanistan with a supportive husband, children, and a home. Her family of origin moved across the country to Maryland, but that would not be an issue, she thought, because "I have my husband now." During her deployment, she received divorce papers from her husband. He was having an affair and waited until she was gone to tell her. He was also moving, as the house was in his name. The deployment became combatting her spouse on custody of their children. She would receive split custody, only after she returning back to the United States. She returned home and checked into a motel. Now homeless, she scouted various friends in the area who had an empty couch for her to sleep. She now craved the experience of Afghanistan and all the consistency it provided. When she was at her lowest point, she intentionally overdosed on medication and was admitted to the hospital. After her suicide attempt, she slowly put her life back together. She held this memory when the people in San Diego learned she was a veteran, she was either told how brave she was or they commented about how she could ever leave her kids to deploy.

U.S. ARMY SERGEANT ANDREW LLOYD

Sergeant Lloyd enlisted in the U.S. Army after high school, and although his enlistment time line maps perfectly with the September 11, 2001, terrorist attacks, he knew in high school he wanted to enlist. For as long as he could remember, he wanted to be in law enforcement, and viewed a position as a military police (MP) officer as a way to jumpstart that career as well as pay for college. 9/11 is an interesting bookend for Andrew's service; it was certainly a factor in reinforcing his decision to enlist, yet he also tells the story with some unease. He did not know what it was that military police officers "actually did down range in Iraq," so it is only his hindsight looking back on his eighteen-year-old self that he describes that initial sense of patriotism.

When Andrew graduated from high school, he left for basic training. Soon after his advanced training course ended, his MP company was mobilized to Washington, D.C., to provide security at the Pentagon. During this mobilization he heard rumors that his MP company was headed for Iraq. Andrew knew that he was going too; it was not that he was afraid but that he was "dealing with the reality" of what this meant. That reality of what they were embarking upon was not lost on him. Each day as they would leave their barracks en route to the Pentagon, they would pass Arlington National Cemetery and "take in all of the graves." This transfixed him. Combat, in his mind, would change him, but in ways he could not have imagined.

He arrived in Iraq in 2004 and spent a year living in austere and rural conditions. This deployment was, for all intents and purposes, "pretty normal, physically speaking." Andrew, in passing, comments that he almost died a couple times, their outpost was attacked by mortar fire often, but, as he noted, it was a pretty normal experience looking back. Although a friend of his was killed, they came home with other wounds. However, when discussing the "horrific" April 2004, he paused to reflect on the service members killed in action (KIA) throughout Iraq and truly what a "terrifying time" it actually was. He spoke often of the "luck" that followed him, whether it was driving over an improvised explosive device (IED) for four months before they noticed it and blew it in place (BIP). After one patrol, he found a bullet buried in the armor of the HUMVEE. Andrew finally stopped to realize that "they were trying to kill *me*."

At this point, Andrew fast forwards his memory to his next Operation Iraqi Freedom (OIF) deployment to support the surge in 2007. This deployment would lengthen to fifteen months. He considers himself fortunate that these deployments were spread out to almost two years. Most of his colleagues deployed for twelve months, came home for twelve months, and deployed again. Eight months into this tour he received the news that his unit would be extended an additional three months; a twelve-month deployment became fifteen. This was completely "demoralizing." The extension exposed various coping strategies, both healthy and those that are a hindrance, among the MP company. Whether watching videos of animals dying, or replaying videos of actual combat, Andrew noticed that his battle buddies were barely hanging on. Andrew was also able to comment that most of his "bad habits" would be waiting for him back home. What happens when service members are more accustomed to combat than home?

DEATH BY A THOUSAND CUTS: COMBAT, HYPERVIGILANCE, AND HARASSMENT

The deployment experiences of my participants are varied in degree, yet contain enough overlap to draw some conclusions. First, yes, there are clear inciting moments of a morally injurious event. The quintessential life-or-death moments in which service members make split-second decisions and from those liminal spaces must now reckon with the consequences. However, other narratives represent more the ongoing, daily, traumatic reminder of being at war. Lisa and Angela's experience of this ongoing traumatic experiences involve military sexual trauma (MST).

As Lisa thinks about her service, and particularly her season in Afghanistan, there are moments of trauma that forces her to "confront the notion that the 'sacrifice' they endured (and continue to endure) serves no greater good and is, in fact, disruptive to any benevolent outcome."[2] It is crucial that caregivers recognize that inciting incidents of course take place, but we also need to affirm the traumatic experiences of ongoing assault and harassment. Lisa details a season of life in Afghanistan in which she was routinely harassed and assaulted. Whether it was waiting in line to order her food in the DFAC and feeling the hands of camouflaged men grope her; or the sexual quid pro quo her former First Sergeant propositioned before deciding whether to approve her leave for a leadership course: her time in Afghanistan was chaotic and not *just* because of the Taliban.

Lisa recounts this story with immense ambiguity. Her First Sergeant was her leader, her role model as an aspiring NCO, and also an individual invested in her safety. Before they deployed, he gave her a knife as an additional weapon, and taught her how to do the most damage if the Taliban tried to hurt her. The cognitive dissonance is palpable; it was not the Taliban that she needed the knife for, but rather the very First Sergeant who promised to take care of his soldiers. For Phillip and Andrew, their experience includes the destruction and recklessness of their actions in combat.

As a truck driver within a Forward Support Company (FSC) element, Phillip's primary mission was driving support vehicles to outlying combat outposts (COPs) and forward operating bases (FOBs) throughout western and southwestern Afghanistan. Typically, in the wars in Iraq and Afghanistan, a FOB was a larger base that served as a hub for various units (American and coalition forces). A COP, on the other hand, typically served as a "fire base" that targeted a specific area of enemy activity. Smaller elements lived on a COP. Therefore, a COP is reliant on resupply and support from larger FOBs in the area. Phillip's missions could be quick "turn and burns," in which his convoy would drive to a COP, drop off or pick up the needed equipment, and return to the larger FOB. Other missions, were more detailed, and necessitated two to three days "outside the wire."

Phillip's MIE occurred early in his deployment. While driving the lead truck of a thirteen-vehicle convoy on a narrow Afghanistan road en route to a FOB in southwestern Afghanistan, Phillip's convoy approached a small, beat-up white sedan driving in the opposite direction. Phillip saw the vehicle approaching as they both came upon a narrow bridge. Phillip described the scene as follows:

> In our pre-convoy brief, we had been warned about an IED threat. Or maybe it was a VBIED [vehicle-born IED] . . . I don't even know, there was always something briefed to us that was "last seen on such and such route"! The weird thing about that day is that traffic usually stops for us. We have this power when we drive over there. . . . We had power over there. . . . It felt good at the time! Stay away from us! When we were out rolling, we would move people out of the way, whether by blaring our horns or just riding someone's bumper until they moved. This damn car didn't move, though, as I approached it. Like five minutes before,

a bus was getting way close to us. It hit my side mirror! So, I am already amped and ready for anything. I see the oncoming car. . . .

The standard operating procedure (SOP) for the convoy dictated that, as the lead vehicle, Phillip should continue driving at his current speed for the safety of the convoy. Speed is a constant companion in the narratives of combat. Andrew also noted how fast they moved, and this tendency for speed followed him home, but more on that momentarily. Phillip was one of thirteen total vehicles in that convoy, each vehicle containing at least two soldiers and as many as eight. As he approached the car, in his mind he urged the car to stop—almost begging it to. In his headset he could hear his lieutenant, Chavez, commanding him forward: "SPC Campbell, are you kidding me, drive! Don't slow down!" Phillip double-checked that this was the correct course of action. "Are you sure, sir?" Phillip responded, and it was confirmed that he was to continue driving. Phillip's hands began to sweat, and his internal temperature elevated as he approached the vehicle. He could feel the sweat dripping down his back, and he used his green army-issued leather gloves to wipe his brow.

Phillip made eye contact with the driver, a middle-aged man with deep brown eyes dressed in traditional Afghan attire. Phillip's truck struck the oncoming car at a choke point as the convoy approached the bridge. His truck smashed the car, forcing it off the road and into a stone embankment. The sound of metal crunching and glass shattering drowned out the commands coming over the headset. Just as loud as the crash was his own heartbeat racing in his chest. He quickly looked in his side mirror, which had been damaged in the earlier incident with the bus, but through the cracked glass he did not see anyone exit the vehicle; he merely saw shards of glass puddled in a pool of coolant, oil, and other fluids. Phillip's heart sank. Lieutenant Chavez called over the radio the familiar call sign, "Charlie Mike" (for *C-M*), which all the soldiers understood meant "continue mission."

Now, at home five years later, he thinks about that man and that incident. What did it do to the driver? Was he injured, or worse, dead? Were there other passengers in the vehicle he could not see? How did this affect the driver's livelihood? Convoys have ways to mitigate such risks, whether by signaling with a siren (which Phillip's truck had, but did not use) or by firing a laser beam meant to disorient individuals, but not harm them.[3] The same procedures used when deciding to fire upon an enemy combatant were used for convoy SOPs. Phillip's convoy that day either did not have them fully in place or did not utilize them.

One thread that ties the morally injurious events together is power. Phillip's sense of power in Afghanistan ("We have this power when we drive over there. . . . We had power over there") was completely turned upside down when he returned home and was no longer within a military cultural construct in which he could impart force on communities. Therefore, what was "right" within the moral orienting system of a military cultural context, i.e., making the decision to crush the vehicle, is not "right" in his own personal morality. Further, and essential for reintegration purposes, such a dilemma is not as easy as just realizing one is driving on an American highway. Andrew comments on power as well. For similar reasons as Phillip, to come back to the United States with a sense of superiority; whether driving fast on highways, or losing your temper with civilians. Speed and tenacity were survival techniques Andrew internalized during his tours in Iraq.

The power does not easily dissipate, and when we have it, it can be difficult to let go. Phillip and Andrew both still think about driving with a similar mentality. Phillip pre-checks his truck daily, maps his route, maps an alternative route, and maintains laser-focused situational awareness of his environment. His current convoys typically end with a family dinner. When he and his daughter were on their way to dinner one evening (their own two-vehicle convoy), Phillip approached a traffic light turning red. He instinctively accelerated into the intersection and stopped his truck, blocking oncoming traffic to ensure that his daughter made it through. He is proud that his instincts got everyone to dinner safely. He laughs about it, and says, "my wife thought I was nuts." However, he understood it as "leading my daughter to safety." It is noteworthy that this memory is from 2018, not from a time soon after he returned home in 2013.

Before deploying, driving was Phillip's stress reliever (and a reason he decided to be an 88M). He would hop in his truck, turn on some heavy metal music (usually Metallica) and drive the long way home from work. Now, driving is not only a reminder of combat, but also a reminder of how high the stakes can get while driving. Today, driving is a source of anxiety and physiological arousal. At its worst, Phillip had physical sensations in his neck until he felt sick to his stomach. Phillip refused to drive for an entire month after his redeployment because he could not separate civilian life from deployed life. This stress while driving was one of the first indicators that Phillip recognized that told him he needed to see a counselor for help, but he did not do so for a few more months. He described how his anxiety from driving prevents him from doing basic tasks. Because he avoids driving, he ends up avoiding the outside world. The entire driving experience is what troubles him: from checking the vehicle, to driving, to the hypervigilance of driving.

Lisa, on the other hand, was on the receiving end of someone else's abuse of power. The perpetual harassment and assault wore her down. She recounted her powerlessness in terms of who would believe her? In a culture, in her experience, of victim blaming, she knew that if she were to come forward, she was jeopardizing her career or further reprisal. The people in leadership were the ones betraying her.

Lisa's and Phillip's MIE are consistent with Shay's organizational understanding of moral injury outlined in chapters 1 and 2. To recap, Shay argues that for an MIE to occur, three things all need to happen: (1) there has to be a betrayal of what is morally right, (2) by someone who holds legitimate authority, and (3) in a high-stakes situation.[4] Phillip felt betrayed by his command during an incident that he now must reckon with as a betrayal of what he believes is right. The entire opening chapter of Shay's first text, *Achilles in Vietnam*, centers on "bad leadership" as "a cause of combat trauma."[5] Shay argues that Homer's *Iliad* is a narrative of coming to terms with failed leadership. The connection, for Shay, comes from taking the Greek word *thémis*, which can mean *character*, *fairness*, *loyalty*, or *morality*, and expanding its meaning, through his analysis of the Homeric narratives, to include "what's right."[6]

Shay specifically unpacks his concept of moral injury as a soldier's "fiduciary assumption."[7] Shay connects a modern soldier's dependence on other soldiers as somewhat anathema to the soldiers of the Homeric narratives. The Homeric soldier, in Shay's estimation, depended on himself and not on "a chain of people he would never meet."[8] In opposition to this, Shay describes the Vietnam experience as follows:

> The vast and distant military and civilian structure that provides a modern soldier with his orders, arms, ammunition, food, water, information, training, and fire support is ultimately a moral structure, a *fiduciary*, a trustee holding the life and safety of that soldier. The need for an intact moral world increases with every added coil of a soldier's moral dependency on others. The vulnerability of the soldier's moral world has vastly increased in three millennia.[9]

Both veterans, I think, would agree. Lisa, dependent on her First Sergeant for mentorship and counsel; Phillip, as a truck driver, was vulnerable and dependent on intelligence gathering for the safety of his route. Phillip depended on a lieutenant who not only knew the SOPs, but also knew mitigating factors when things went awry. Phillip was not angry with his lieutenant though; Phillip saw the leader's action for what it was: a split-second decision that Phillip must now process. Lisa, from a more systemic angle, was vulnerable and dependent on an organization to not reduce her to her sexuality.

In chapter 2, I noted the juridical-clinical framework that Wiinikka-Lydon unpacks. Within that discourse, there is a useful trajectory within critical theory and philosophy to understand the dehumanizing aspect of power. Philosopher J. M. Bernstein writes about the dehumanizing nature of sexual harassment, sexual assault, and rape. Bernstein connects this to moral injury due to the broken social contract between humanity. One's dignity has been taken away, not only by the perpetrator, but further by the severed trust with one's comrades. Lisa is right to not only view her 1SG as a threat, but view the entire unit through the lens of betrayal. Speaking to the communal aspect of dignity, Bernstein offers, "the dignity of the person, or in an older jargon, someone's having a soul, is a social construction; the dignity of the person just is what comes to be through the forms of recognition through which the intact, self-moving body comes to be: the dignity of the self is the reflective articulation of the moral integrity of the body."[10]

Rarely in my interviews did each veteran speak about systemic failures of the US military in the Global War on Terror. Andrew came the closest as he talked about the absurdity of the mission in Iraq and an MP's role in the surge. Further, Andrew is the farthest removed from his enlistment. He views his service from a more antiwar perspective. For him, this realization was an unfolding; there was no single event of realization. It, importantly, happened in a community of veterans and concerned citizens. Andrew has experienced the potential that a community of veterans and civilians can have to come together to hear these stories.

For those who are not from the same political perspective as Andrew, as a researcher, I was curious if any sort of critique of the system would emerge. In my estimation and experience, it is precisely the system that warrants a critique. Phillip's unit, for example, had mitigating procedures that would have prevented the entire incident, but they were not utilized. Upon reflection, I believe Phillip was not hiding his critique. In chapters 5 and 6, I note that, within the liberative praxis methods of this project, this is precisely an opportunity for military chaplains to function as Gramscian intellectuals supporting counterhegemonic communities. Intellectuals lend their intellectual resources, when appropriate, to strengthen the counterhegemonic cause.

At this point, a caregiving caveat is necessary. It would not be advisable for military chaplains to impose their own political critiques onto service members and veterans, just as they would not impose their own religious commitments. The mission of the counterhegemonic community is to offer a community that veterans and civilians are not finding in other spaces. This community, however, does maintain a critical lens.

Speaking to this interaction with Phillip by way of example, he was told he and his unit were some of the last service members to be in Afghanistan. However, when the interviews took place in 2018, the U.S. military was still in Afghanistan, and the Taliban had more territory than at any other point in the war. Shay's organizational analysis of moral injury can be levied against the entire military-industrial complex. Service members continue to die in Iraq, Afghanistan, and myriad other countries with no end in sight; this is a moral injury—a lack of clear mission is its own deep betrayal. Within liberative praxis, there is a responsibility to present the realities of these critiques. Now, if I were to impose my systemic critique on Phillip's narrative—without an invitation—I would be deploying violence onto his story. All this is to say that there is an inherent tension when considering the most liberative way to critique systems from within counterhegemonic communities, a tension that I attempt to assuage in chapter 6.

As we turn to the themes of reintegration, it could be useful to remember Shay's statement that "moral injury is an essential part of any combat trauma that leads to lifelong psychological injury. Veterans can usually recover from horror, fear, and grief once they return to civilian life, so long as 'what's right' has not also been violated."[11] The experiences in this chapter became a moral injury when, "the primary intensity of moral stress exceeds the person's capacity to remain grounded and whole."[12] Phillip knew at the time that what he did was wrong, but he recounted the event in such a way that he could not have acted any differently. As Lisa stood at parade rest, she knew that this was a horrific betrayal. As Andrew watched the deterioration of his MP company as twelve months seeped into fifteen. However, it was only when they each reintegrated post-deployment and had other bouts of post-traumatic stress that the full weight of these incidents set in. Through hours of interviewing, coding, and analyzing, three distinct themes of alienation emerged: belonging, the isolation and difficulty of reintegrating within the context of a civilian and military divide, and a deep theological betrayal. I will address them in turn.

AN ANALYSIS OF THE ALIENATION OF REINTEGRATION

Belonging: Coming Home

In different, yet congruent ways, each veteran returned home to an anticlimactic reality, and a reality in which "somebody turned the volume way down," as Andrew noted. When Phillip returned home after his deployment, he expected a celebration. More specifically, he expected a parade with balloons and city leadership to be at the unit welcoming him home. However, this was not the case. There was a small reception at the unit, but it was haphazard and thrown together. As the chartered bus approached the unit, there were families and friends holding signs: "We missed you, Daddy!" Welcome home, Mommy!" "Sergeant Gonzalez is our hero!" The unit, though, were scrambling to get the grill working. The festivities, which everyone had known about for months, had somehow fallen through the cracks. The battalion commander was not present at the reception. The only soldiers present, who were not members of the deployed force, were a handful of full-time soldiers who worked at the unit. Phillip had envisioned how this day would go, but it was turning into a disaster.

Lisa knew there would be no welcome-home hug for her either. The North Carolina heat reminded her of the intensity of Afghanistan, yet she traded kinetic chaos for "freedom"—whatever that meant. Instead of joining welcome-home parties, she retreated to the barracks and cleaned. Armed with a bottle of bleach, she scrubbed the entire bay. Similarly, Angela landed back in San Diego, and went to check into a motel. She was now not only a combat veteran, but she was also newly divorced and homeless. Each veteran hoped and yearned for connection, but this was not the case for their story.

Each veteran went on leave and the reservists I interviewed each took some time off from their responsibilities as reservists. However, when they returned to their weekend drill two months later, they were surprised how much had changed. The unit had, for all intents and purposes, moved on. In their absence, the unit had to fill slots. This is a normal phenomenon of a military unit: the mission goes on even at home station (the unit at home station is often referred to as "rear detachment"), and therefore people are needed in each slot. When the deploying unit returns home, there is some tension with regards to who gets the responsibility. Phillip noted that "the unit treated us like nothing, I didn't feel like I mattered. It all just felt different."

Phillip recalls a new soldier, probably around nineteen years old, who was getting to the unit after his training at Fort Jackson. Phillip pulled him aside, and like a potential mentor offered his support. "I want to teach you what it is really like. How we really do things over there." The young soldier looked at Phillip with rejection; at least that is how Phillip internalized it. He thanked Phillip, but he knew what he was doing. Phillip simultaneously felt his heart sink and his disappointment and anger rise. What was his purpose now? Phillip was quick to clarify that new soldiers were not necessarily the problem. He was perfectly able to work with new soldiers and train them based on his experience. However, what he found were new soldiers that didn't care that Phillip had deployed—they didn't ask for his experience. It was at this point that Phillip noted that his service in Afghanistan did not matter. The atmosphere was one in which every soldier was on the same level; Phillip was just as competent as that brand-new soldier fresh out of Advanced Individual Training. He experienced this as a failure of leadership: the officers and noncommissioned officers (NCOs) did not do enough to properly reintegrate the veterans back into the unit. There was no recognition of what these individuals did in Afghanistan. To this, Phillip noted:

> You are only used and are here for a reason for a certain season. Once your time is done, *you* are done. All right, thank you, that's it! You are left with all these emotions and experiences that you bring home and what do I do with that? You

know, what do you do with it? You get back [to the unit] and there is just . . . you know . . . you are just expected to get back into "drill mode." You are lumped in with everyone—even those who didn't deploy. They don't know anything. We were thrown back to the mix. They should have used our experience, but we were just like everyone else. . . .

This season of Lisa's life was filled with disillusionment: the institution that previously provided so much of her identity was a source of her alienation. Her desire to attend a leadership course was thwarted when she refused to sleep with her First Sergeant. The ideals of that recruiter who came to her high school seem like a fading memory. Lisa processed her reality through writing. During our interviews she returned to her journals for the first time in seven years:

We're going home. My God, we are actually going home. It blows my mind that in five days I will come back to the United States and life will resume "as normal." What does normal look like anymore? I have all of these memories and experiences and regrets and dreams. Real life scares me. In a way, I liked my life on hold. No bills to pay, no laundry and formations and paperwork and school. It is all a little daunting now. I want to cry but I've forgotten how. I'm torn in a thousand directions, my mind, my feelings, everything. I'm exhausted in every measurable and non-measurable way.

There is just so much to do, so much I want to be, but it seems as though I always run out of steam. I don't want to see my family and I don't want to deal with the pain or the "combat stress" or the hard things. I've done enough hard things this year. I'm tired of hurting. I want to bracket the entire world and run way for a while so I can breathe. I want wide open spaces and no expectations and to be in the arms of someone who loves me despite my flaws.

She had not opened her journal until we spoke. When she talked about *that* Lisa, she is proud of her younger self. She sees a young woman with grit and resilience; she got through.

Once the veterans were out of the unit and the military, there were times when they would get nostalgic about service; yet, it was more complicated than that. Andrew knew it was time to get out. He notes that he was "dealing with my PTSD before I knew I had PTSD." He shares a vulnerable story about a layover in the Dallas airport on his way back to Iraq in which he stopped into the USO. While watching the time pass, he decided to record himself reading a children's book that he could send his nine-month-old son. Throughout his retelling, he stops multiple times to tell me how lucky he was to be stateside, not out of a sense of guilt, but truly a gratitude that he returns to throughout our time. He forces himself, through tears, to read *Green Eggs and Ham*. What sort of dad was he becoming, he wondered.

At times, Phillip does miss his "combat family," and he certainly misses the feeling of being needed.[13] However, for all the times that Phillip had reflections like this, he had just as many that centered on his family at home and how he would never want to go back to Afghanistan. I asked Phillip what I can see in retrospect was a leading question: whether he would go back to Afghanistan if the opportunity presented itself. After a long pause, he replied, "No way. It is a young man's game. I have my family and have too much to lose." He reached for his wallet and pulled out a picture of his granddaughter, at this point around eighteen months old, and with tears in his eyes, said, "I need to be here for her. Someone needs to watch *Frozen* with her."

Lisa, after signing out of her unit, intuitively *knew* she needed help: "will they give me the little blue and green pills that promise to keep a handle on the fog—give me windshield wipers for my mind and calm the nervous thrumming of anxiety that has replaced a steady heartbeat inside of my chest? I need an electric shock. Restart this mind and heart." However, when she got to the Behavioral Health Specialist during her out-processing checklist, she noted she was "fine." Angela, navigating similar behavioral health hurdles, noted she was fine. All she wanted was to go home—whatever that meant for her now. There is so much tension and redefinition of what *home* means in her story. She laments that she got to a point in which she attempted death by suicide. Everything she knew, both before the chaotic experiences of combat, was erased; she did not recognize the coordinates on this new version of the map of her life.

Phillip felt alienated not only from his unit but also from the society around him. I asked Phillip at the end of our third interview if he were to give advice to a reserve service member transitioning to civilian life, in terms of expectations, what he would say. He replied:

> They don't care about your service—some may thank you. But, truly, they don't care about your service. Kids are hyped to kill like *American Sniper*, and want to know what combat and killing are like. I work with a guy who is a disabled vet and another coworker said, "So, I don't care. Why should I care? You all volunteered for it!" I wanted to punch him in his face! To serve in the military is not an easy task! I am with my third therapist because of how hard this shit is! No, they don't care.

Within that complex paragraph is exactly what drives this project: there is a level of dissonance that Phillip experienced with civilians *and* the military. They may thank him for his service, and he feels some gratitude for their words, but it is their actions that reveal their true (lack of) appreciation. Phillip did not know this coworker, but he had conversations with him in his head: "How dare you! You have no idea what it was like!" Phillip was instantly transported back to the roads of Afghanistan, and his MIE. He often goes back to this memory. He wakes when nightmares of that day creep into his sleep. In a cold sweat he jerks up and looks around his bedroom. The man in the car with the deep brown eyes is peering into his soul. In the nightmare he hears a cry, which sounds like a small child. He distinctly remembers not seeing anyone exit the vehicle, but now he cannot remember what he did not see. He does not trust his memory. Now, though, he wonders: Did I really volunteer for *that*?

When Andrew got out of the Army, he decided to continue working in law enforcement. He thought he was handling his post-deployment life with composure and resilience. However, the salve he had carefully applied to his combat trauma came back when he started to respond to mass shootings. It became "the war coming home." Stepping across the threshold of a school filled him with a fear that he had not felt since Iraq. The fear came from a sense that he was *not* being prepared. To counteract this feeling, he spent $6,000 of his own money on medical supplies and weapons upgrades. However, holding his modified assault rifle, he knew deep down that if the day should come and he was not ready he would never forgive himself.

Society: Divided Identities

The theme of "divided identity" connected most directly to the ideological critique I am advocating. I want to specifically handle the ideological implications of reintegration as I have not yet discussed the direct role ideology and society played in shaping each interviewed veterans' reintegration. Phillip's alienation from the army introduced a unique experience of being a reservist and attempting to reintegrate a MIE. After the deployment, the individuals that perhaps understood Phillip the best "went their separate ways." His mentor, Sergeant First Class Eric Clausen, a blond surfer from San Diego who joined the military at eighteen to travel the country, was direct, tough, but always dependable. He taught Phillip more efficient ways to clean his truck; he taught Phillip easier ways to get Wi-Fi access in remote Afghanistan; and, he taught Phillip how to be a deployed husband. Phillip wanted to *be a* SFC Clausen to the new soldier he met when he returned to drill. When the unit returned to the United States, Clausen transferred, but did not say good-bye. Transferring is not atypical for a restructuring unit: reservists went back to school, took new jobs, or simply just went away for a while to reconnect. Phillip was left, alone, to reintegrate into civilian life. This reintegration was difficult and is still ongoing.

Phillip is torn when he discusses society. On one hand, he is appreciative of most people who thank him. Remember, above, he stated, "Some people thank me, and I thank them for their appreciation." He wants to believe that people are appreciative of his service and sacrifice, but with some groups of civilians, their actions tell him otherwise. Phillip can feel it when ideology is functioning. The handshakes are flimsy. The eye contact is nonexistent. Phillip believes that his generation is "the last one to get a thank you." Gratitude, for Phillip, is complex. Ungratefulness, on the other hand, is a simple act of dishonor. Phillip recounts one story he "can't get over":

> I had these two conversations with people that I can't get over. This one dude told me, "we shouldn't even be there. You all are fighting a lost cause." I was so pissed, what has that guy ever done?! I do work with a lot of vets, and they get it. One former Navy SEAL told me to keep my head down. His last words to me were, "don't get killed." He was the coolest dude. He was a Vietnam vet, and he was called a "baby killer" when he came home. He told me, "These people don't realize that I killed somebody yesterday, and could kill you today and not even blink an eye." These people are so lucky. He did the job. I was just a reservist. . . Guys like that helped me get through.
>
> I met this woman; it was around Christmas time, as I was walking into the mall. I was wearing a unit T-shirt from Afghanistan. She said, "When did you serve?" I explained that I just got back, and am in the Reserves. She looked confused and said, "we send the Reserves to war?" What the hell was that?!

Even in that vignette, Phillip is quick to dismiss his own service as not as valiant or as worthy of gratitude as a Navy SEAL from Vietnam. In both situations, he feels himself to be, less than, and others convey minimization as well. He is not a Navy SEAL, and he is not active duty. His part-time status in the Reserve component separates him. From these interactions, Phillip decided that it is better to internalize his service, and not share it. The people he meets either do not understand the complexity of military service, or only want to hear about the special operators (the "Operators," as they are colloquially referred) because these operators are the ones the movies are about. Operators killed Osama Bin Laden. Operators have television shows (*The Unit* or *SEAL Team*) and movies (*Zero Dark Thirty*, *Lone Survivor*, or *American Sniper*) made about them. Operators are elite.

Phillip returns in his mind to Afghanistan: leading convoys on Afghanistan highways. He recalls one specific mission in which they came upon an IED. This was not a test. Phillip hears in his headset, "I got something back here. I see wires!" Phillip recalls experiencing tunnel vision. He wanted to collapse into himself, and hide. He knew, though, that he had a responsibility to the convoy. The SOP dictated that they radio the EOD team (Explosive Ordinance Disposal). What the SOP did not dictate was how to handle the boredom of waiting hours in the heat for EOD to arrive. Phillip "kept his cool" and told those around him to "keep your head on a swivel." Phillip was mentally alert and attuned to everything around him. He noticed a flicker of light out of his peripheral vision, and quickly got his rifle ready. He was relieved when it was merely a young woman walking with her mother. EOD finally came and casually walked up to the IED site and blew it in place (BIP). The convoy was "Charlie Mike." Phillip shared numerous stories like this, in attempt to show that in Afghanistan—in combat—it does not matter what type of unit or from what component you deploy: war is war.

These exchanges also highlight the importance of perspective when Phillip discusses his service. The perspective spoke to two aspects in particular: first, Phillip carried a new personal understanding of perspective in civilian life. He understood that "all this could be taken away." Second, he still dismisses his service as not as worthy of appreciation as that of the Navy SEAL. Phillip appreciates that there are others who were called on to *do* even more.

One of most poignant reintegration experiences for Phillip happened at a mall, as it so often does for a reintegrating veteran. The mall is the battleground for the military and civilian divide. Former Marine, Phil Klay describes the mall as:

> Back home was shopping malls and strip clubs. Over here was death and violence and hope and despair. Back home was fast food and high-fructose corn syrup. Over here, we had bodies flooding the rivers of Iraq until people claimed it changed the taste of the fish. Back home they had aisles filled wall to wall with toothpaste, shaving cream, deodorant, and body spray. Over here, sweating under the desert sun, we smelled terrible. We were at war, they were at the mall. ... There's something bizarre about being a veteran of a war that doesn't end, in a country that doesn't pay attention.[14]

Following the conversation with the woman above, Phillip continues shopping for Christmas presents and describes American children pleading with their parents for toys—pulling on their parent's arms and legs for more. "Can I have that, Mom? Please!" "You said if I was good Santa would bring this game!" Phillip was surrounded by the deafening sound of kids screaming about what toys they wanted. He then shows me pictures of children in Afghanistan, barefoot, that he met when he was outside the wire. Deafening screaming to a peaceful photograph. He remembers this one boy throwing rocks. He became angry with the American kids and laments to me vicariously: "You have nothing to complain about. I am more thankful and grateful now. The fathers are out in the field working just so the kids can eat dinner that night." Just as Phillip felt isolated from reservists in his unit, Phillip now felt isolated from Americans when he went shopping.

The last element of this theme that is instructive and speaks to the ambiguity of these wars in Iraq and Afghanistan is how to handle veteran reintegration when the sheer number of veterans increases. Phillip can admit and emphasize that that civilians just do not know what to say—even after almost twenty years of war. What comes out instead is, "Wow, you must have had it rough" or "Thank you for your service." Phillip felt those ideological sentiments as empty signifiers that did not mean anything. Andrew calls the gesture a token. In his mind, society does not really think about the wars in Iraq and Afghanistan because we have been fighting for too long.

In the midst of not feeling understood within society, Phillip remained special to his family. He seldom talked about how difficult it was reintegrating with his children; however, his marriage was the source of much strife. From his family's perspective, not much changed initially while Phillip was deployed. A critical thing to realize about the modern Iraq and Afghanistan deployment is the level of comfort and predictability soldiers experience; they are always connected to their families. Phillip's FOB had Wi-Fi; he could message his family at any hour of the day. This, of course, is a powerful asset for service members to connect quickly, yet it also has severe implications. One implication is trying to communicate the daily reality of what combat encompasses. The deployed service member will typically *always* have the story to tell that is "more pressing" or "more important" than the spouse or loved one back home. An IED attack always takes precedence over a long line at the post office. Because of that, Phillip ignored his family, or at least he tried.

There is a latent conflict: the people he wants to understand him cannot because he is not entirely willing to divulge his complete narrative. He would send Facebook messages or texts, but they never knew when he went on missions. When he returned, his family was eager to catch up. His family was ecstatic to see him, but Phillip felt isolated, as "they didn't get it." The family wanted to celebrate in public, at a bar or restaurant, but he did not want to. The anxiety of being around people became too much. The once-social Phillip was no longer social. Phillip had changed.

Phillip's newfound intuition to stay home and not go out and visit with friends and family became too much of a stressor on his marriage. Phillip started drinking more heavily. He would watch his unit's deployment DVD, drink whiskey, and replay "what ifs": What if the Taliban had overrun the base? What if the Afghan National Police had opened fire? What if he had hit an IED? The drinking, the isolation, and the changing temperament were too much for Vicki, and she told Phillip she was moving out. He looks down at his hands, rubbing them together calmly, and simply states, "She said I changed." He became more forgetful. He was not attentive to her needs. He is perplexed about this as well; this is not the Phillip he knows, or wants to be. He used to be a man who "went with life more. I didn't sweat the small stuff." Those four months from Vicki apart tested Phillip's resolve perhaps more than the deployment. He haphazardly described it as:

> Those four months killed me. I lost a ton of weight. The not-knowing. . . . I couldn't sleep. I felt sick all the time. I didn't know how to express myself. She wasn't listening to me. I felt like my feelings didn't matter. Everything was "you, you, you"—what about my feelings? I finally said to her, "just come home." She did, and we worked on things. I do snap sometimes when she asks questions! Why do I have to answer everything—she is not my mom! I try not to sweat the small stuff.

Once Vicki returned, Phillip started to take his reintegration seriously. This would include three VA therapists, medication for a new PTSD diagnosis, and new patterns of thinking. Phillip knows that he is a different man than before he deployed. Now, though, he wants Vicki to know that he "goes to the VA and realize that this is your husband now. He is here working on it. I am working. My family is everything, and I am not going to lose them again. Although the doctor said I will have it [PTSD] forever. You just learn to deal with it. I thought I was doing great. It just kept getting worse."

Andrew, while trying to volunteer in a law enforcement capacity, ends up living in his mother's basement. He is recently divorced after an affair, an affair he says was his way of dealing with his PTSD. From his mother's basement, his view of himself shifts: what was once a warrior is now terrified that he will not be prepared to protect the most vulnerable. His mother ends up kicking him out of her home. In hindsight, he understands her reasoning. Through this jolt he begins to see a therapist. "I got to tell you, Maggie saved my life."

Angela, inch by inch, puts her life back together. A job at Starbucks gets her on her feet and enables her to eventually get another job *with a desk*; the desk was symbolic. She, in her estimation, had status again. She had a place to belong. Each individual took steps to work on their most intimate relationships. For some, however, there is a theological difficulty with which to contend.

MY GOD, MY GOD WHY HAVE YOU FORSAKEN ME?

For Phillip and Lisa, an alienation from God during reintegration surprised them the most. They experienced a deep internal theological change when they returned home. Phillip felt this alienation as betrayal. To understand this betrayal and its shift, his spirituality before is important to understand. During Basic Training, Phillip could not get enough of the Bible. He received a New International Version (NIV) Bible in Basic Training that he still uses to this day, and poignantly, that he took on missions in Afghanistan. The spine is worn, the pages are filled with highlights and scribbled notes, and there are numerous bookmarks of favorite passages—what Phillip called "life verses." Some of his life verses connected military topics and his Christian faith, specifically a sacrificial and redemptive narrative of military service. For example, he would read about the attributes of a warrior and the similarities with the faithfulness required of being a Christian. He gravitated toward the Christian trope of sacrifice found particularly in the Gospel according to John 15:13: "No one has greater love than this, to lay down one's life for one's friends." This sort of hermeneutic is not solely Phillip's, as it is prevalent in many Christian congregations and in the US civic religion of militarism. Professor Kelly Denton-Borhaug astutely notes that the "redemptive interpretation of soldier's suffering is a kind of thinking that is alive and active in American civil religion, as well as in religious institutions across the nation."[15] Phillip saw connection and overlap between the soldier he was becoming and the Christian he was becoming.

After his deployment, Phillip went through a deep season of isolation and abandonment, and it is noteworthy that this was also the season in which he felt isolated from God. Without his "battle buddies" co-suffering with him, his faith also went through a season of isolation. Phillip's faith, then, is a communal faith. This is not to say that Phillip's personal spiritual disciplines were reserved for a communal setting; rather, Phillip's prayer life, daily Bible devotional reading, and listening to praise music were daily personal disciplines that were sustained alongside a committed community. Within his platoon, a small group of soldiers became "accountability partners" and prayed for one another. When other soldiers had weekend passes, Phillip got his Bible out and read through the New Testament.

He was confident that God would always guide him. Phillip relied on—and felt like the prophet—Jeremiah, in which God would provide a plan of welfare and hope (referencing the prophet Jeremiah in Jeremiah 29:11); Phillip was proud that he, like Isaiah, stood up and was sent into harm's way (referencing the prophet Isaiah in Isaiah 6:8); finally, Phillip had been confident that God would strengthen him to do all things (referencing Saint Paul from Philippians 4:13). These texts and others functioned as "life verses" for what he was doing in the Army: God empowers warriors. That God, however, stays behind in Afghanistan. With more confusion and pregnant pauses, he eventually describes his pain:

> I am still trying to figure it out. I don't know. I had the chapel services and Bible studies. I don't have a connection anymore. I felt, almost, thrown into the world over there. When I think about it now I still just get more confused.

> I still try to pray every day and I thank God for my life—and my grandbaby. That is a *real* joy in my life. I can thank God for that. I still, though, don't know about my daily life and forget that God is here with me now. There is something about how serious everything was in Afghanistan, with faith, with family, with my own life.

I followed this up with, "was God with you in Afghanistan?"

> Yes! I felt strong because of him. I had people around me who held me up. People held me up. Little things like prayers before missions always helped me feel ready. I would leave the gate knowing that God was with me. In the middle of a mission—like when we broke down in the middle of a valley—I didn't necessarily feel abandoned. When I got home, though, I felt abandoned by everything.

During reintegration, Phillip and Lisa describe God as distant. Perhaps it is more apropos to describe this distance emerging *in* Afghanistan, but Phillip, in particular, had to come to terms with this God's absence once the alienation surrounded him during his reintegration. Phillip did not feel like he had a prosperous hope (returning to Jeremiah), and he certainly did not feel as though he could do all things (returning to Philippians). Phillip had dedicated his life to Jesus and had a community of believers around him, but this did not prevent terrible things from happening.

Phillip continues to visualize his MIE and, in particular, the man's eyes. He thought his deployment was necessary to help the people of Afghanistan, because that was a dominant ideology we all were immersed within, but all around him he saw fatigue, corruption, and death. Phillip needs a God that could not only reckon with that reality, but a God that was present in that experience and in his reintegration experiences. This vision of God is possible, and as Dietrich Bonhoeffer quipped: "Only the suffering God can help."[16]

RELIGIOUS RESOURCES

I want to transition to the religious resources that sustained Lisa and Phillip and their deployment in Afghanistan: prayer, a religious community, and music.

Prayer and Meditative Spaces

Phillip's prayer life after returning from his deployment was another indication that something was off. Prayer used to be natural—almost second nature—to Phillip. Now, it was reminiscent of his entire alienated faith journey. Phillip described this: "When I think about faith and prayer, I am still confused. My faith was done. Why did God send me there? I was so strong before. Why is it gone now?" Phillip could not commune with God because he felt as though the connection was severed. More than severed, though, Phillip's spiritual practices of prayer were mired within his inability to concentrate. Phillip notes that a side effect of his PTSD is "I have a hard time thinking. I get caught up in my talking. Even right now. It is hard for me to speak. I lose my train of thought. So, I start praying, but then forget what I am praying about." It became easier to stop altogether as he did not have people in his life to walk through enacting new practices of mindfulness or centering himself to pray again. When I asked Phillip if he wanted to continue praying, he answered desperately, "of course!" Phillip is still looking for connection to other people who are either a) Christians who "get it" or b) veterans who "get it," and if those two could coalesce that would be preferred.

Lisa connected most with the Zen Buddhist practice of sesshin. She spent two weeks in a dojo under the direction of a roshi (also a healthcare chaplain). These meditative spaces became a way for her to quiet her mind and the memories of Afghanistan. What emerged in this practice was an individually taxing experience of sitting for hours and clearing her mind; however, it was done *in* community with other practitioners. Even though the "community" looked different than a traditional group therapy session, each sesshin provided her a space to work out—and clear out—her process. The intense working lowered her vulnerabilities and inhibitions about concealing her pain. Her days were spent with intense manual labor, such as chopping wood, with relatively few hours of sleep. She described the period as a way to bring "Zen into action." The community was responsible for all her sustenance, and all she was required to do was work and meditate.

Religious Community: The Search for Solidarity

When Phillip returned home, he felt isolated from his spiritual battle buddies who had accompanied him through some of most trying times of his life. Andrew noted the constant monotony of meeting people who lived lives with no "sense of mission." People at Phillip's local church just could not compare. I have already detailed the importance of chapel in Phillip's life. Chapel was a constant religious experience in which he felt nourished, welcomed, and challenged. He held the various military chaplains up in admiration; these men and women were Phillip's heroes. Even in my interactions with Phillip, I felt an instant rapport when he learned I, too, had been to Afghanistan—and as a chaplain. Chaplains lived a similar deployment experience as him. They could understand what he was experiencing.

When Phillip attempted to reintegrate into American Christian life he was surprised how much had changed for him. The sermons seemed shallow. The pastors preached about money—*a lot*. Importantly, the pastors were not military chaplains. As they were not chaplains, Phillip was not willing to follow them, or even take their advice. A natural by-product of feeling alienated is not reversing it: Phillip no longer wanted to spend time with people who did not get it. The entire experience left him searching for what he had in Afghanistan. This is a journey he is still on. Phillip's reintegration and spiritual interventions are still communally possible. In chapters 5 and 6 I note that within the liberative praxis methods of this project this is precisely an opportunity for military chaplains to function as Gramscian intellectuals supporting counterhegemonic communities. It is from within the veteran community that support and intervention can thrive.

Before Lisa underwent her season of sesshin, she spent time "floating," waiting for a community to normalize and affirm her. She hid and negotiated the space of her reintegration alone. She hints at moments of connection. She mentions a woman, a civilian, who started a conversation with "I know you have been through a lot, but if you don't want to talk about it, that's completely fine too." She exhaled and sighed that this was a space in which no expectations were levied against either. She describes this conversation as "the warmth that awaits you when you step outside." However, the family members who noted she was different and asked her how many people she killed in Afghanistan were "stepping outside in the rain and cold . . . and you just don't want to enter into that."

Music

I saved music for the end, because music is the space in which Phillip experiences not only his truest self, but an experience that also includes God. Phillip had always been into heavy metal. He played the guitar, and enjoyed heavy rock music with good guitar riffs. He had been in bands throughout his youth and late adolescence, and even "jammed" with other adults as he grew older. When he became a Christian he thought that season of his life was over. He assumed that Christian music would be more akin to hymns than to Metallica. However, he found Christian heavy metal. In our first interview, Phillip brought up how important the band Kutless was for him during his deployment. He describes that his morning routine *always* included "blasting Kutless." He had the music loaded on his phone and would listen while working out and when he had downtime during a convoy. Kutless enabled Phillip to have a religious connection while also staying true to his passion for loud guitars and heavier, faster music. A previous unit left an old acoustic guitar in Phillip's living area, and he would pick it up from time to time and play Kutless songs.

Later in my interviews with Phillip, I asked more specifically what it was about Kutless that spoke to him. I was curious whether or not certain albums influenced him, or if other bands had a similar gravitas in his life. Kutless was a band he shared with his twenty-two-year-old daughter. He got emotional describing driving her to school listening to two Kutless albums: 2009's *It Is Well* and 2014's *Glory*. She would hop into his pickup truck and immediately turn on Kutless. Phillip's responsibilities were to roll down the windows and get ready to sing along. With the windows down, the brisk autumn air rolling in, they would sing together—as loud as they could—and smile at each other after each song.

After one drive, she told him she wanted to get baptized. They had been going to church as a family for about a year. The pastor asked if Phillip wanted to participate in her baptism, and he was ecstatic. They chose a song from Kutless' 2009 album *It is Well* as a dedication song in her baptismal service. *It is Well*, is a worship album with classic worship hymns remixed and sped up with detuned guitars and fast drums. When he first heard the album, he could not believe the power it provided. During that baptism he felt closer to God than he had previously. Kutless "had everything for me: family, God, and heavy."

When he got home, though, music was no longer an asset. While talking about Kutless, Phillip said, "let me tell you a story":

> I didn't play my guitar for three years after I got home. I was numb... Like my hands too... I had no creativity. Three years. My old drummer that I used to jam with before deployment called me up. I had been avoiding him, but he kept calling. It was the medicine I needed. We jammed, and like never stopped. The emotion was so strong in me that when we stopped playing I started crying. It was a release. This is it. I couldn't find myself when I came home. I lived a life in Afghanistan that I didn't know if I was coming home. Would I make it out alive? That little deployment did a lot of changes in my life—I didn't think it would, but it did.

I asked Phillip if he thought music was waiting for him. With tears in his eyes, he said, "It was waiting for me." The complexity of Phillip's reintegration experience provides a canvas for further theoretical exploration. Phillip's alienation and subjectification within a totalizing neoliberal hegemony demand a serious account. Through a revised praxis method of correlation, and Lartey's liberative praxis, chapter 5 now asks new questions, posits answers to some of Phillip's questions, and uses Phillip's questions to reform

IMPLICATIONS FOR PRACTICAL THEOLOGY AND THIS PROJECT

I close this chapter by elaborating on a plea I laid out in chapter 1. To properly reintegrate Phillip, Lisa, Angela, and Andrew into society, I proposed, two steps: first, they need a space in which they can fully articulate and "re-author"[17] their MIEs, and second, military chaplains (primarily) and practical and pastoral theologians (secondarily) must critique the ideology of American military service to create space for proper reintegration. I also mentioned in chapter 1 that greater collaboration is needed with civilian clergy. For all intents and purposes, the participants do not talk to their former battalion chaplains from combat anymore. Once they all return home, after their season together, each go their separate ways. How, then, can military chaplains—functioning as Gramscian intellectuals—creatively assist in preparing the veteran, the religious community, *and* civilians for a veteran's homecoming?

Ultimately, if this critique of how cultural ideologies can amplify an existing moral injury is not properly dealt with, Phillip will continue to receive a hollow "you must have had it rough"; Lisa will hear "you look different; the life is gone from your eyes"; Angela will be reminded that civilians could never *do* what she did; or the liturgical platitude of "thank you for your service" while nothing changes at a cultural level. The veterans I spent time with experience those platitudes, and it did little to reintegrate them back into society. It is time to move forward in our practical theological method toward developing normative ways to reflect theologically within cultural studies' use of ideology.

NOTES

1. Osmer, *Practical Theology*, 4.
2. Kristen J. Leslie, "Betrayal by Friendly Fire," in *War and Moral Injury: A Reader*, ed. Robert Emmet Meagher and Douglas A. Pryer (Eugene, OR: Cascade Books, 2018), 248.
3. Former army chaplain David W. Peters describes the usage of these laser pointers as a way of warding "off approaching vehicles. . . . The Iraqi drivers believe they are lasers from weapon scopes or something. . . . Cars pull over to the side of the road while their drivers give us dirty looks." "Sin Eater," in *War and Moral Injury:*

A Reader, ed. Robert Emmet Meagher and Douglas A. Pryer (Eugene, OR: Cascade Books, 2018), 214.
4. Shay, "Moral Injury," *Psychoanalytic Psychology*, 183.
5. Shay, *Achilles in Vietnam*, 196.
6. Ibid., 5.
7. Ibid., 14.
8. Ibid.
9. Ibid., 15.
10. Bernstein, "Suffering Injustice," 313.
11. Shay, *Achilles in Vietnam*, 20.
12. Cynda Hylton Rusthon et al., "Invisible Moral Wounds of the COVID-19 Pandemic: Are We Experiencing Moral Injury?," *AACN Advanced Critical Care* 32, no. 1 (2021): 120.
13. M. Jan Holton, *Longing for Home: Forced Displacement and the Postures of Hospitality* (New Haven, CT: Yale University Press, 2016), 131.
14. Phil Klay, "The Warrior at the Mall," *New York Times*, April 14, 2018, https://www.nytimes.com/2018/04/14/opinion/sunday/the-warrior-at-the-mall.html.
15. Denton-Borhaug, "Like Acid Seeping into Your Soul," 117.
16. Bonhoeffer, *Letters and Papers*, 361.
17. White, *Maps of Narrative Practice*, 61.

Chapter 5

The Centrality of Community in Moral Injury Support:

Theological and Cultural Studies Analysis

This chapter operationalizes the latter tasks of Lartey's liberative praxis outlined in chapter 3. Throughout this chapter I will work the experiences of my participants (Lartey's first task of liberative praxis), using these liberative praxis and practical theological methods with a theological and sociological conception of community from Dietrich Bonhoeffer. To better interpret the data from chapter 4, this chapter is broken into two parts: in Part 1, I more formally return to Stuart Hall's analysis of ideology. This theoretical background unpacks how Hall's ideological analysis assists in rethinking—and ultimately opposing via spiritual practices—dominant cultural ideologies surrounding military service. After a theoretical introduction, the interviewed veterans' questions and concerns are brought more formally into the dialogue. Their experiences assist in identifying how media imaging impacted his MIE and its reintegration.

Part 2 attends to the theological side of interpretation, namely an examination of Bonhoeffer's sociological ecclesiology as a lens to better conceptualize the liberative praxis functions of the communal-contextual spiritual care paradigm. Here, I will focus primarily on Bonhoeffer's first published work, *Sanctorum Communio: A Theological Study of the Sociology of the Church*. To better facilitate our conversations, I will start to put together the pieces of how Hall's "oppositional" reading of ideological texts is not only possible, but provides a pathway toward healing via counterhegemonic communities.

Part 1. Return to Ideology

In their interactions with civilians and some active duty military personnel, the veterans herein understood that their military service was something *other*. Of the three dominant themes unpacked in chapter 4, a divided identity within society is most prescient at present. My return to Hall's ideology is strengthened with the analysis of my participants in mind. Phillip and Lisa both knew that they did not compare with the "hero" elite "operators" valorized by civilians, and therefore Phillip's questioning, and his reintegration, critiques a dominant ideological production of these wars in Iraq and Afghanistan. Lisa described a lack of representation in the way her service was described. As a young enlisted woman who dealt with ongoing sexual abuse and harassment, the narrative of war did not encompass her narrative. As she reminded me, "war is told from the perspective of a white man, and I do not see myself very often in those stories." To drive home the point about ideology, though, in chapter 4 Phillip detailed an interaction in which he came to the realization,

> They don't care about your service—some may thank you. But, truly, they don't care about your service. Kids are hyped to kill like *American Sniper*, and want to know what combat and killing are like. I work with a guy who is a disabled vet and another coworker said, "So, I don't care. Why should I care? You all volunteered for it!" I wanted to punch him in his face! To serve in the military is not an easy task! I am with my third therapist because of how hard this shit is! No, they don't care.

Phillip and Lisa were encountering ideology. Young men, in particular, want to enlist to become like their heroes, such as Navy SEAL operator Chris Kyle, protagonist in *American Sniper*. In chapter 2's genealogy from Marx's ideology critique through Althusser and Gramsci I highlighted one of my two primary cultural studies conversation partner: Stuart Hall. I have only provided a brief note about his biography and *theorizing*, as it is now appropriate to explore his work on ideology at greater detail.

OPPOSITIONAL READINGS AS RESISTANCE

Hall, contra Marx, posits that cultural forms (e.g., the media or advertising) are not simply reducible to a superstructure related to an ideologically driven economic base. Rather, for Hall, culture represents constitutive elements of society as a whole. Hall eliminates the rigidity that Marx and later Marxists posit between the base-superstructure paradigm.

As noted in chapter 2, for Hall, ideology is not a false consciousness, but rather, a socially constructed site of struggle. Therefore, as I started to unpack in chapter 2, all that can be done is choosing how to function within ideologies. The question becomes *which* ideological frameworks we engage—rather than focusing on how to escape them. Ideology is "a function of the discourse and of the logic of social processes, rather than an intention of the agent."[1] Ideology cannot be explained, but remains partly unconscious.

Hall, as with Gramsci, insists that ideology resides in both language and the common sense of the culture. Further, through a thorough analysis of hegemony one can locate how vectors of power attain dominance. One space in which common sense is produced is within mass media. The media "serve, in societies like ours, ceaselessly to perform the critical ideological work of 'classifying out the world' within the discourse of the dominant ideologies."[2]

The goal, then, is to examine how hegemonic forces ideologically shape media to maintain common sense consent. I will discuss how ideology generates certain types of veterans in the wars in Iraq and Afghanistan through the hegemonic "readings" of war films. Hall's analysis and usage of the media offers a practical and tangible way to understand, not only hegemony, but also discursive practices in spite of hegemony. For Hall, media classifies the world within four ideological roles. First, media participates in producing social knowledge. Media, then, have "colonized the cultural and ideological sphere."[3] In their colonization, media creates images that groups and classes can project their lives onto. In other words, media's imaging and social knowledge constructs for groups an image of the world, or *lived* realities of groups, that is the world.

Second, media reflect on the plurality of the different worlds. The plurality comes into play as the media is tasked with filing and mapping the variety of life (this is as simple as region, classes, cities, neighborhoods, or minority interest groups) as there does not exist a universal ideological discourse. Media, then, looks to reflect on the "inventory of the lexicons, life-styles, and ideologies which are objectified there."[4] Third, similar to Gramsci, media is tasked with producing consensus and consent. Media can hold minority opinions and views that produce an illusion of objectivity. In this third role, Hall notes that ideology functions to produce "consensus, the construction of legitimacy—not so much the finished article itself, but the whole process of argument, exchange, debate, consultation and speculation by which it emerges."[5]

Fourth, and vitally for this project, the media operates *independently* of the state. However, this does not mean that dominant state ideologies do not creep into media's imaging. Hall's point, though, is that "counter-vailing" voices can emerge. Events that the media report on do not *just happen*; they, through the media, are "made intelligible."[6] For Hall, events are made intelligible through encoding, as the process by which a preferred ideological reading follows. There are limits to interpretation in how these products are encoded. The producers cannot have a "perfectly transparent communication" in how a message is decoded, and this is why the lived experience of the individual decoding a product is critical for interpretation.[7] The "repertoire" of codes and the application of meaning are from dominant ideologies. This is not necessarily a conscious decision, but is from a stream of meaning. It is, however, because the stream of meaning is within common sense and consensus, they appear as though they are the only available meaning. Hall, thus, often refers to ideology as unconscious. Reporters, or other "professionals" within mass media, act as if they are pursuing the values of their institution; however, they are actually functioning within a dominant ideology. Where are the sites of resistance within this consensus and dominant ideology?

Encoding/Decoding

People decode from their social locations (Hall's "plurality" mentioned above). Hall's critique of Althusser's view, for example, is that it fails to account for ways in which *both* ends of the production (encoding *and* decoding) distort meaning. Hall critiques linear models of communication theory that progress from sender to receiver via a message. Importantly, the linear model fails to account for the social and political implications of discourse. In its place, Hall favors a circuit. Hall emphasizes the activity, rather than passivity, of discourse.

Therefore, ideology is also in the product's decoding. It is precisely this potentiality in which holistically reintegrating veterans with moral injury *is* possible. Decoding, then, is the result of a disrupted variety of textual readings. Hall defines decoding as, "before this message can have an 'effect' (however defined), satisfy a 'need' or be put to a 'use,' it must first be appropriated as a meaningful discourse and be meaningfully decoded."[8] The process of decoding is not as simple as unpacking encoded messaging; rather, Hall proposes three "hypothetical" reading positions of decoding's construction: "dominant-hegemonic," "negotiated," and "oppositional."[9]

First, in the dominant-hegemonic, Hall, utilizing his Gramscianism, describes how the media texts of a culture are infused with the ideology of a dominant group. This is a tangible example of how Gramscian consent affects everyday life. The messaging is decoded exactly how the encoded message was intended. These viewers are "operating inside the dominant code," and according to Gramsci's common sense.[10] The producers of these messages operate by a "professional code" of experts.[11]

The second reading position is a negotiated code. Simply put, "majority audiences probably understand quite adequately what has been dominantly defined and professionally signified."[12] These readings highlight contradiction and, "the legitimacy of the hegemonic definitions to make the grand significations (abstract), while, at a more restricted, situational (situated) level, it makes its own ground rules—it operates with exceptions."[13] Finally, Hall offers the possibility of a counterhegemonic and contrarian decoding of texts, which opposes the dominant-hegemonic position. These counterhegemonic positions of resistance are crucial, and they are imperative for this project. Poignantly, Hall states, "he/she detotalizes the message in the preferred code in order to retotalize the message within some alternative framework of reference."[14]

The Dominant Ideologies of *American Sniper*

Returning more specifically to the reintegration experience of the wars in Iraq and Afghanistan, let's highlight Clint Eastwood's *American Sniper* (2014) as a vignette. Remember, Phillip specifically recounted stories in which special operation service members ("operators") were used as the veteran par excellence during his reintegration experience, so it is prudent to run *American Sniper* through Hall's analysis. *American Sniper* is the story of the protagonist, Navy SEAL Chris Kyle—the "legend" of the Iraq war—becoming the "most lethal sniper in US military history."[15] As a Navy SEAL, and therefore an operator, Kyle represented a type of veteran (a hero, from Moon's typology) with which Phillip was compared. Phillip inherently understood a dominant ideology of films such as *American Sniper*—or at least its effects.

I am not interested in critiquing Eastwood's film *as* film; rather, what I am interested in is what is happening at the level of ideological production that drew an estimated $517 million worldwide.[16] Other films portray moral ambiguity in a more realistic and sobering manner (For example, Ang Lee's 2016 *Billy Lynn's Long Halftime Walk*), but those films simply "don't sell." Suffice it to say, there is something about *American Sniper*, whether that's the ideological message it portrays, the ideological values it espouses, or the ideological patriotism it demands, that nudge one to take a closer look.

American Sniper tapped into aspects of war and collective memory that the public desired. This is Hall's "encoding." Speaking to that, Hall states, "'mass culture' is art that is machine-produced according to a formula."[17] Perhaps it is better to delineate *American Sniper* within the genre of western due to the blatantly obvious gunslinger killing "bad guys." In the *New York Times* review of *American Sniper*, critic, A. O. Scott states that it upholds a Hollywood tradition of the "binary western film," and protagonist Chris Kyle's character throughout *American Sniper* is a binary (gunslinging) Navy SEAL who "despises the savage, despicable evil. That's what we were fighting in Iraq."[18] *American Sniper* attempts an apolitical stance toward the Iraq war and is an "expression of nostalgia" for President George W. Bush's foreign policy.[19]

All this to say, *American Sniper* is an ideological film. As an ideologically encoded film, there is not much room for ambiguity. Michel Foucault speaks to this overall problem of creating binary one-dimensional heroes while not detailing the entire experience when he states, "the problem's not the hero, but the struggle. Can you make a film about a struggle without going through the traditional process of creating heroes?"[20] Lisa noted this when she commented that the female perspective of these wars and the LGBTQIA+ perspective of these wars are still not represented. Her experiences in Afghanistan would entail a ninety-minute film in which she navigates harassment and assault from fellow comrades throughout her tour.

Perhaps one could argue that *American Sniper* contains that struggle and the very resistance that I am advocating; namely, the traumatic aftereffects of killing confronted Kyle. There could be an argument that people are talking about the war because of Eastwood's film. Eastwood himself can even argue that both *American Sniper* and he are anti-war, which he does in the aforementioned *Guardian* article, but I still contend that it is *precisely* in the encoding of *American Sniper* as an easy formulaic entertainment product of the culture industry that obfuscates the actual trauma of these wars, and exists as an ideological prop for the military-industrial complex. For example, the fact that Kyle struggled with reintegrating at home is a reason he is a hero and deserves recognition. *That* sacrifice is part of the "formula" that Hall addresses; it is encoded in the film. It represents the "repertoire" of codes within the film. That script, however, misses the unmitigated decoding of the invisible wounds of these wars. Simply put, *American Sniper* is still too clean and resolves too easily.

American Sniper opens in Iraq during the middle of Chris Kyle's deployment. Positioned as the sniper overwatch, Kyle assists a Marine unit patrolling the area below him. Our view, as people watching the film, is through Kyle's sniper rifle scope. While scanning, he picks up a young boy walking alongside a woman (presumably the boy's mother). Kyle—and the viewer—witnesses the woman hand the boy a grenade. He calls up on his radio to receive confirmation of the grenade before firing, which he does not receive. If Kyle deems the boy a credible threat, he is authorized to shoot. As the boy, with the grenade in hand, begins to run, Kyle pulls the trigger. The victim of this shot, though, is a deer as the film utilizes a flashback scene with Kyle's father in rural Texas. Viewers are left on the edge of our proverbial seats.

The film returns to this combat scene twenty-five minutes later, with Kyle shooting the child and the woman who picks up the grenade after the boy's death. When Kyle is back in his barracks, he receives a word of encouragement that if he hadn't killed the boy and woman that many Marines could have died and that he was "doing his job" to which Kyle states, "Yeah, but I killed him. That is not how I imagined my first one going down." Later, Kyle recalls that memory, as:

> She'd set a grenade. I didn't realize it at first. "Looks yellow," I told the chief, describing what I saw as he watched himself. She's got a grenade! Shit! Take a shot! But . . . I hesitated. . . . I pushed my finger against the trigger. The bullet leapt out. I shot. The grenade dropped. I fired again as the grenade blew up. . . . I truly, deeply hated the evil that woman possessed. I hate it to this day.[21]

How did it come to pass that society did not talk about the implications of an ongoing war in which women and children are in the crosshairs of snipers who *need* to decide whether they live or die? Let's explore, as a discursive tactic, Hall's hypothetical decoding readings of *American Sniper*.

First, the dominant-hegemonic reading is that Kyle is a hero, and our ongoing conflict in the Middle East is a righteous battle. Therefore, in the decoding of the dominant ideological message of *American Sniper*, it became apparent that one could not say a contrary word about the film, because (seemingly) to speak against the film (or Kyle), was to speak against military bodies. For example, when actor Seth Rogen tweeted, "*American Sniper* kind of reminds me of the movie that's showing in the third act of *Inglorious Basterds*,"[22] he was referring to the mini-film within Quentin Tarantino's (2009) *Inglorious Basterds*, entitled *Stolz der Nation*, in which protagonist and Nazi sniper Frederick Zoller kills numerous Allied soldiers in an almost comical propaganda piece for the Nazis (utilizing the exact same binary western style of differentiating "good guys" from "bad guys"). Rogen had challenged the one RSA that one cannot challenge: the military—and especially the service members who sacrificed.

Activist filmmaker Michael Moore also spoke out against *American Sniper* via Twitter stating, "My uncle [*sic*] killed by sniper in WW2. We were taught snipers were cowards. Will shoot u [*sic*] in the back. Snipers aren't heroes.

And invaders r [*sic*] worse."[23] Individuals on social media and conservative cable news immediately retorted to Rogen and Moore that *they* should be killed. Fiftieth Speaker of the United States House of Representatives Newt Gingrich responded to Moore, "Michael Moore should spend a few week with ISIS and Boko Haram. Then he might appreciate *American Sniper*. I am proud of our defenders."[24]

Second, in terms of a negotiated reading, I would argue that Rogen's actual position—not the backlash—represented a negotiated reading. In his tweet was an acceptance of the act of war, but with an inverted twist. Nothing in his tweet called the military-industrial complex into question, and his subsequent apologies confirm this. Rogen recanted his tweet, apologizing, and stating that he was not critiquing the service members, just the Tarantino film. In other words, Rogen was able to acknowledge, "the legitimacy of the hegemonic definitions to make the grand significations (abstract), while, at a more restricted, situational (situated) level, it makes its own ground rules— it operates with exceptions."[25] Rogen's negotiated reading is riddled with contradictions.

Finally, for an oppositional reading and the counterhegemony it offers in resistance, I will dislodge the media reaction, and instead focus on realities immersed in my participants' lived experience. The goal of an oppositional and counterhegemonic reading is not to remove Kyle's service (The goal is more akin to pastoral theologian Ryan LaMothe's "unconventional warriors" and remain "attached to the warrior ethos, but are critical toward a government that uses its military to further the aims of political and economic elites."[26]) but to complicate our complicit acceptance of dominant ideologies and war. The entire film of *American Sniper*, and the vignette above in particular demand a reading with the lived experience of those, like Phillip, Lisa, Andrew, and Angela, who also suffered in combat. This oppositional reading must privilege the voices of those traumatized. With Hall's oppositional reading in place, albeit tentatively, I want to propose spiritual care and counseling postures and practices that fit together with these cultural studies theoretical commitments and are upheld through the participants' remaining chapter 4 themes: belonging and spiritual abandonment.

Part 2. A Communal-Contextual Commitment to Moral Injury Support

COMMUNITIES EXISTING FOR VETERANS

Utilizing Bonhoeffer and *Sanctorum Communio* (arguably Bonhoeffer's "most complex and most demanding book"[27]) is instructive in three unique ways. First, pragmatically, Bonhoeffer offers a lens with which to read liberative praxis and the liberation theologies that undergird that praxis. As one of the primary audiences for this book are military chaplains, reading Bonhoeffer as an ecclesiological guide is more amenable to the culture of the multivalent voices within military chaplaincy, over and against a perception of Marxist ideas within liberation theologies.[28]

Bonhoeffer predates Latin American liberation theologies by, at most, forty-one years if we accept that the *official* genesis of Latin American liberation theology was the 1968 Second Conference of Latin American Bishops (CELAM) at Medellín, Colombia, as I suggested in chapter 3, as Bonhoeffer completed *Sanctorum Communio* in 1927. This is not to say that Bonhoeffer was the originator of liberation theologies, but only that his work anticipated themes and motifs that would shape these theologies. Central liberative motifs that would become paramount for liberation theologies were present throughout Bonhoeffer's work, including: God's preferential option for (and solidarity with) the marginalized, a praxis-based ecclesiology, and a God that is experienced within history—to include experiencing suffering.

It is Bonhoeffer's later work, though, that signaled a liberative theology. In the prologue of *Letters and Papers from Prison*, "After Ten Years," signaling the ushering in of 1943, Bonhoeffer concludes this letter with a paragraph on "the view from below." In this, he concludes, "we have for once learnt to see the great events of world history from below, from the perspective of the outcast, the suspects, the maltreated, the powerless, the oppressed, the reviled—in short, from the perspective of those who suffer."[29] This is fundamentally a statement of liberation theology. Comparing this statement to Juan Luis Segundo, whom I closely unpacked in chapter 3, the inroads are striking. Commitment to a marginalized community, in Segundo and Bonhoeffer, is *the* first step, and theology and reflection always follow lived experience. Segundo's hermeneutical circle comes to some semblances of truth only through practice and commitment. As Segundo noted, "the truth is truth only when it serves as the basis for truly human attitudes."[30]

Second, Bonhoeffer cultivates formative commitments for his life's work within his early writings. The concept of sociality (*Sozialität*) that Bonhoeffer develops and nuances throughout *Sanctorum Communio* represents a road map for where Bonhoeffer not only takes his academic thought, but also how Bonhoeffer embodied his theory in his actions. Bonhoeffer scholar Clifford Green notes that without grappling with the "autobiographical dimension," one's understanding of Bonhoeffer's "theological development would be darkened by inner obscurity or externally imposed speculation, or both"; and more poignantly, Bonhoeffer's "theology was an impulse to his action."[31]

Elaborating on this last point, Bonhoeffer's context does align to the experience of reintegrating moral injury. In his posthumously published *Ethics*, he wrestles with the moral injuries that were waiting for him in his dilemma whether to participate in the assassination attempt on Adolf Hitler. Within *Ethics*, Bonhoeffer begins to discuss the moral responsibility to "act" even when those actions are morally "wrong." He states, "the 'world' is thus the *domain of concrete responsibility* that is given to us in and through Jesus Christ."[32] Importantly, for military chaplains, Bonhoeffer seemingly understands the ambiguity within moral decision-making. When split-second decisions have lifelong implications, Bonhoeffer would perhaps remind veterans, "Responsible action must decide not simply between right and wrong, good and evil, but between right and right, wrong and wrong."[33]

The context of war, morality, and crisis enabled him to disseminate a liberative vision of God that spoke to and *from* the crisis. This is a vision of God that moral injury support needs. Phillip, in particular, spoke of a God, and community, that abandoned him when he returned home. Bonhoeffer rehabilitates a Christology and anthropology that speaks life-giving words of hope to returning veterans. Christ is present in each experience with the veteran. Therefore, when Phillip asks why God has abandoned him, a Bonhoefferian Christology allows a framework to shift the conversation.

Third, and finally, focusing on *Sanctorum Communio* is instructive in an attempt to rehabilitate its complexity and application within the corpus of Bonhoeffer's oeuvre. The early theological work of Bonhoeffer, spanning from 1927 to 1933, is typically viewed as a period to merely get past to get to the "proper" theological texts, such as *Discipleship*, *Ethics*, or *Letters and Papers from Prison*. This hermeneutical move, in which scholars judge the early work on the notoriety of the latter work, misses a deep complexity within *Sanctorum Communio*. Theologian John Phillips, perhaps unwittingly, makes this argument as he notes that Bonhoeffer needed to "break free from his intractable ecclesiological theory" found in *Sanctorum Communio*.[34] However, to call *Sanctorum Communio* merely an ecclesiological study is to miss the nuance of what Bonhoeffer is developing. In the Preface he offers this as his précis:

> The more this investigation has considered the significance of the sociological category for theology, the more clearly has emerged the social intention of all the basic Christian concepts. "Person", "primal state", "sin", and "revelation" can be fully comprehended only in reference to sociality.[35]

Bonhoeffer's sociality offers the nuance and setting for his ecclesiology. Essentially, it is Bonhoeffer's Christocentric ecclesiology, the "Christ existing as church-community" (*Christus als Gemeinde existierend*), in which people learn to be in community for other people that enables me to circle back to pastoral theology and the communal-contextual paradigm as a way to holistically care for the veteran community and society at large.[36]

DIETRICH BONHOEFFER'S THEOLOGY OF SOCIALITY

Dietrich Bonhoeffer was born on February 4, 1906, in Breslau, Germany (present-day Poland). Dietrich was one of eight children. Dietrich's parents, Karl and Paula, raised him and his siblings, Karl-Friedrich (born 1899), Walter (1899), Klaus (1901), Ursula (1902), Christel (1903), Sabine (Dietrich's twin, 1906), and Susanne (1909) in a middle-class family. Dietrich's father, Karl, was a distinguished professor of neurology and psychiatry while also serving (from 1904 to 1912) as director of University Hospital for Nervous Disease in Breslau until leaving for the most prestigious appointment in Germany: the University of Berlin. Various academics, scholars, teachers from numerous facilities, and other intellectuals frequented the Bonhoeffer household. Lifelong confidant and student of Bonhoeffer Eberhard Bethge described Dietrich's upbringing as within "a family that derived real education not from school, but from a deeply-rooted sense of being guardians of a great historical heritage and intellectual tradition."[37] This intellectual tradition would be instrumental in Dietrich's upbringing and their collective resistance to both

World Wars—especially after Walter Bonhoeffer, a soldier in the German army, died near the end of World War I, in April 1918.

Dietrich's decision to go into ministry and study theology was met with disappointment and some ridicule. For the Bonhoeffers, the church represented "a poor, feeble, boring, petty bourgeois institution."[38] The pushback only strengthened Bonhoeffer's resolve to reform the institution. He would begin his formal theological education at Tübingen University in 1923. Dietrich transferred, in 1924, to the University of Berlin to focus on his studies under the country's renowned theological professors, to include professor emeritus of church history Adolf von Harnack. In 1927, for his dissertation, however, Bonhoeffer decided to write under the tutelage of Reinhold Seeberg. This enabled Dietrich to write a "half-historical and half-systematic work."[39]

More specifically for the data in this project, Bonhoeffer worked against a prevailing Enlightenment (and German idealism) notion of the human as an autonomous individual. Bonhoeffer argued that humanity becomes human with the recognition of our responsibility to humanity. This interdependence of humanity is also central to what it means to be a Christian. There are myriad gaps in this abridged biography of Bonhoeffer's early life, and no mention of where Bonhoeffer's conspirator activist theology of resistance would take him once Adolf Hitler becomes German Chancellor in 1933. This is naturally a problem of scope: this is not a book on Bonhoeffer, but rather a focus on his view of sociality as it is experienced in community. However, as I mentioned above, the autobiographical nature of Bonhoeffer's academic work found ways of expression in his personal life. I will offer one example before transitioning to *Sanctorum Communio*.

In 1935, Bonhoeffer took a formal role with the Confessing Church (*Bekennende Kirche*), teaching at their Preachers' Seminary in Zingst, which later moved to Finkenwalde. The Confessing Church's goal was to train as many pastors as they could before the Nazi Gestapo ultimately shut them down at the end of 1937 (after being declared illegal in December 1935). Bonhoeffer in these brief two years cultivated a community, his own sanctorum communio. Once the Gestapo closed the seminary, Bonhoeffer rushed to compile the theology of this community, and *Life Together* (*Gemeinsames Leben*) was published. *Life Together* is a blueprint for working out sociality. The seminary was more closely aligned to an intentional community than a traditional academic environment. The students were immersed in life together, while at the same time highlighting a rhythm of solitude as well. Daily life was filled with eating meals together, daily prayer, sermon practice, lectures, and bible study. The practice of intercessory prayer, and being available for one another in prayer, signaled a return to the theology of *Sanctorum Communio*, to which I now turn.

The Community of Saints

Throughout *Sanctorum Communio,* Bonhoeffer filters the majority of theological doctrine through sociality. Sociality is the antithesis of the Enlightenment understanding of a rational individual operating autonomously from others. Bonhoeffer saw an interconnection in which, "the concepts of person, community, and God are inseparably and essentially interrelated."[40] Sociality, then, is not only the interconnection of humanity but also the interdependence of humanity. His task became one of developing a philosophy and theology that matches this inseparability. I will highlight, from Bonhoeffer's précis, his theological anthropology and his statement on revelation—a revelation that is not dogmatic, but one that is social and *personal* as it is embodied in the programmatic "Christ existing as church-community" (*Christus als Gemeinde existierend*).

Theological Anthropology Reinforcing Belonging

Phillip was dismayed that his army unit, society, and God alienated him. This alienation was cultivated specifically in his experiences of falling through the cracks during reintegration. The army values of duty, selfless service, and leadership were nonexistent. Phillip's understanding of the camaraderie of the military disappeared, and his doubts during reintegration can be directed at Bonhoeffer, and his project in *Sanctorum Communio*. Essentially: how does Bonhoeffer's theological anthropology foster an understanding of not only community as the collection of individuals, but what becomes of the ethical commitments of Bonhoeffer's community?

For Bonhoeffer, the person is both an individual and a corporate entity. Bonhoeffer starts *Sanctorum Communio* unpacking his version of the I-You relationship.[41] The person in Bonhoeffer's work is not merely an individual; the person always exists in a corporate communal setting. The person, for Bonhoeffer, is a relational construct. Humanity does not exist through individual persons but through responsibility for other persons. To expand upon humanity's responsibility for others, Bonhoeffer posits that "the individual exists only in relation to an 'other'; individual does not mean solitary. *On the contrary, for the individual to exist, 'others' must necessarily be there.*"[42] To be a human is to be in relation to others in community.

To be in relation to an other is an encounter with the divine You. "*God or the Holy Spirit joins the concrete You; only through God's active working does the other become a You to me from whom my I arises. In other words, every human You is an image of the divine You.*"[43] The divine, then, constitutes the other as a You. The divine is present in each encounter. The assertion that the divine is present in each moment is a liberative motif. This

God is precisely what Phillip needed upon reintegration. When Phillip got home, he: "felt abandoned by everything"; he "didn't have a connection"; and, he felt "thrown into the world over there." Phillip searched for a God that is with him in his isolation; a God that is with him in his moral injury; a God that is with him during his marital separation. God experiences the pain, suffering, and marginalization of God's people. This does not eradicate God's transcendence.

Bonhoeffer makes the same movements above with respect to humanity existing for another in an individual encounter (The I-You) with fresh social and communal responsibilities as a symbol of revelation. Bonhoeffer takes the emblematic fall of Adam and Jesus' restoration as an archetype, a *Kollektivperson*. Adam was the original "I" before the divine "You." God, through Jesus, reconciles a restored humanity before the divine "You." Bonhoeffer builds this communal argument by differentiating between *Gemeinschaft* (community) and *Gesellschaft* (society). A *Gemeinschaft* is a community, properly understood as a place in which people structure their lives and create meaning—such as, families, a "people group," a nation, friends, and culminating in humanity overall. A *Gesellschaft*, on the other hand, function at a smaller level, a "structure of purpose."[44]

Now my concerns for reintegration are beginning to take shape. The divine's call to support and embrace the other is what wills communities. Individuals within communities have an ethical responsibility for an *other's* You. This ethical imperative comes from the divine:

> God does not desire a history of human beings, but the history of the human *community*. However, God does not want a community that absorbs the individual into itself, but a community of *human beings*. In God's eyes, community and individual exist in the same moment and rest in one another.[45]

Humanity is brought back into solidarity with God through the "vicarious action" (*Stellvertretung*) of Jesus. I now want to spell this out even further in terms of revelation.

Social Revelation: Christ Existing as Community

It is only possible in Bonhoeffer's sociality to say "Christ existing as community" after he articulated Jesus as *Kollektivperson*. Revelation is social (and personal), and it exists in the community. Revelation is a manifestation of divine love. Jesus as *Kollektivperson* represents God's giving of Godself. God's love restores humanity "by revealing God's own love in Christ, by no longer approaching us in demand and summons, purely as You, but instead by *giving God's own self as an I, opening God's own heart*.[46] The vicarious action—*Stellvertretung*—is humanity's call to exist and love one another. Bonhoeffer describes this as *"Church-community and church members being structurally 'with-each-other'* [Miteinander] *as appointed by God and the members' active 'being-for-each-other'* [Füreinander] *and the principle of vicarious representative action* [Stellvertretung]."[47]

Bonhoeffer offers a response to veteran alienation and longing for belonging. With Bonhoeffer's understanding of God's presence *in* the community, and our ethical responsibility to those within our community, the participants could potentially have an ongoing community that he experienced in the army—prior to reintegration. As a chaplain, and one who journeys with service members for short passages of time, I would want the participants—and other veterans—to know that these communities are still not idyllic, and people will forget their story; however, the vision is for a community that can bear the stories of our veterans upon return. Now, *how* communities are to support and live out this vicarious representation falls outside Bonhoeffer, and I turn to the communal-contextual paradigm of spiritual care and counseling for insights.

COMMUNAL-CONTEXTUAL APPLICATION

Moving away from Bonhoeffer, but maintaining his communal focus, what role does spiritual care and counseling have in advocating for veterans within a liberative praxis in which communal solidarity is privileged? Phillip would sit, alone, in his basement with his military memorabilia and remember the people with whom he served. He can still recall the first chapel service he attended at Fort Jackson during basic training. The pews were filled—shoulder to sweaty shoulder—with soldiers who wanted to go to church. Phillip laughs, recounting that it was either church or get assigned extra duty. "Who wouldn't go to church?" Slowly, though, a community emerged. These soldiers were going through similar training exercises as Phillip; these soldiers were at a similar stage in the basic training schedule.

After that initial Sunday service, he was out in the field at Victory Tower, a tower utilized to not only develop the skills of rappelling, but, more important, learning confidence to overcome fears. Phillip was scared. When it was his turn, he peered over the ledge at the seemingly inconceivable height. While situating his feet in a more balanced posture he kicked some loose dirt off the platform and watched it fall. The fall seemed to take forever. Phillip froze. With a similar heart pounding that would return in Afghanistan, Phillip wanted to quit. "I hate heights!" A soldier from a different platoon was already on the ground and looked up and saw his battle buddy, Phillip, struggling: "Come on, man! You got this! We are right here." He began to clap for Phillip. Phillip wiped the sweat that was pooling from under the brim of his helmet, and clumsily rappelled down the wall. When he reached the bottom, the soldier erupted with applause. During chapel, the next week, the soldier addressed this to the group: "You all should have seen Phillip! He was so brave, man!" Phillip holds onto that memory because it was only through the encouragement of a "Christian battle buddy" that he was able to get off the obstacle. As inconsequential as that memory may seem to others, in the grand scheme of Phillip's experiences, it operates perfectly in how Phillip envisions community: people supporting one another to achieve goals that he or she could not complete individually.

Within spiritual care and counseling, the communal-contextual paradigm, just as its name indicates, seeks to reestablish connections to religious communities, both as a source of providing care and as a place of truly building community around one another. The community is a manifestation of God's memory for all people. God is in relationship with human beings by "hearing us, remembering us, and bringing us into relationship with one another."[48] Also implicit in its name, this paradigm is focused on taking contexts seriously as a site for interpretation, and as this project has prioritized thus far, this is in keeping a liberative practical theology.

The paradigm's concepts are built upon the theological reflection that God remembers God's people. John Patton unpacks how this understanding of God and God's mission influences how communities of faith provide care. Phillip's experience in Afghanistan would agree with Patton, while also adding nuance. Phillip *felt* "remembered" in Afghanistan. While waiting on Explosive Ordnance Disposal (EOD) to clear a suspected IED, stranded, in the middle of an exposed valley, Phillip recalls a peace that could only mean God was with him. At home, during reintegration, Phillip did not *feel* anything. How can communities of faith provide care when God is nowhere to be found?

Phillip has experienced the answer, though. He noted that when he scaled Victory Tower, God was with him in the cheers of his battle buddy. In Bonhoeffer's work, this is Christ existing as community. In Bonhoeffer's understanding, humanity does not exist through individual persons but through responsibility for other persons. To be a human is to be in relation to others in community.

From Phillip's experience, and his situational analysis, the communal-contextual paradigm is beneficial. In the communal-contextual paradigm there is an emphasis on involving a community both in providing care "officially" and on including the community in the healing (or what Patton refers to as "re-membering"[49]) process. Remember, for moral injury, any intervention that only focuses on pathologizing and does not include a *community* element for reintegration is destined to come up short. Moral injury is context dependent. This is vital for military chaplains to recognize. There is, of course, an essential place for individual therapy and pastoral counseling in moral injury treatment; however, if we are to broaden our understanding of moral injury away from isolated combat phenomena and to reconceptualize it within community *and ideological* values, then I think this paradigm and its functions of care provide greater opportunities for lasting healing and the implementation of practices of support.

To analyze these systems, feminist pastoral theology is properly situated to critique and formulate new epistemologies for care. Feminist methods afford me a critical consciousness to examine the intersections of power and oppression as they influence Phillip, Lisa, Andrew, and Angela and other veterans. Therefore, I would like to give some attention to feminist concerns as they assist in sharpening reintegration questions. This will also prove beneficial as it exemplifies the concerns of the communal-contextual paradigm.[50]

Feminist and womanist pastoral theologians have pointed out that the epistemologies of women have not been taken seriously in theological discourse. The reverse has consistently been true: the experiences of white men have been upheld as normative and as the ideal. Carroll Watkins Ali has helpfully deconstructed the influence of Seward Hiltner on pastoral theology. Throughout her text *Survival and Liberation: Pastoral Theology in African American Context*, she convincingly argues that Hiltner's method is not sufficient for poor African American women and the communities they represent.[51] Hiltner's functions of care (healing, sustaining, and guiding) do not accurately speak to the communal struggles of African American communities. To speak individualistically in counseling without focusing on what impacts the entire community will not lead to long-term healing. It may present temporary solutions, but not the lasting liberation that Watkins Ali seeks. To embody Watkins Ali's method, the caregiver must acknowledge, "There also needs to be ongoing provisions for persons whose critical needs are extended over a long duration."[52]

Watkins Ali adds three community-based functions to Hiltner's functions: "nurturing" (which Howard Clinebell also addressed), "empowering," and "liberating."[53] To nurture the community, one must have an ongoing commitment to provide care that empowers counselees to have the strength to face various struggles within their community. The function of empowering contains the insistence that the struggle for liberation and emancipation must come from the oppressed people themselves. Within veteran support, then, an empowerment would include not only that communities listen to the narratives of veterans, but also that those communities relinquish and redistribute resources to these veteran voices.

To expand on one small example: sermons on veteran support cannot be relegated to two high holy days of the veteran calendar: Memorial Day and Veterans Day. Empowerment from the veteran community would also include deciding where—and how—to use resources, spend time, and causes the community undertakes. Watkins Ali provides this empowerment in community so the members of the community can resist the systems that oppress. Finally, the liberating function entails praxis. It involves working together as a community to eliminate oppression. The significance of Watkins Ali's work is her insistence that, to adequately care for a community of people, the caregiver must address the systemic forces of oppression that keep the community down.

It is evident from chapter 4 how Watkins Ali offers inroads for addressing my participants' experience. They *had* communal support in Afghanistan, and it is primarily the *lack of* a community during their respective reintegrations that has exacerbated their moral injuries. Phillip would benefit from a nurturing presence to affirm his concerns, while also working in a liberative manner to connect him to other veterans and community members who are aware of the realities of the complications of reintegration. There is one remaining feminist concern that operates as a pivot for this section and names a piece of the collective experience that pervades all others and further complicates dominant ideologies: neoliberal reason systematically oppresses people and leads to suffering. When a veteran either transitions off active duty and begins the process of seeking employment, or when a reserve or national guard veteran returns to civilian life, the platitude of "thank you for your service" is less thankful.

VETERANS AS NEOLIBERAL SUBJECTS

Before transitioning to chapter 6, I want to address an additional theme that emerged in my time with my participants: the *"hegemonic project"* of neoliberal ideology.[54] Through addressing neoliberal reason, I will also expand and deepen a critique in how it functions within veteran reintegration of moral injury. This critique is instructive, as it not only speaks to each participant's experience, but it also forms a bridge from spiritual care and counseling to cultural studies. Through Phillip's exchanges with civilians, his army unit, and even God, he functions within neoliberal reason as a neoliberal subject: what matters is no longer what Phillip as an individual offers society; rather, as a neoliberal subject, Phillip is worth what he can produce through the capital that is his very existence. Phillip was anxious to go back to work; he did not know what a normal job would feel like anymore. During Phillip's first day back, he met with his immediate supervisor. He was surprised—caught off guard even—by his supervisor's attitude. His supervisor welcomed him back by lamenting:

> My boss loses staffing when you guys leave the country. They don't have their people. Soldiers are always gone. It is either your [monthly weekend] Drill, or some schooling, or another deployment. . . . You volunteer for this, and you volunteer for that. . . . My boss is tired of hiring vets. Phillip, just because you're a vet doesn't mean you are a good worker! I have had to fire many of you straight up lazy dudes.

Phillip was dumbfounded, and just nodded his head in agreement. It was during this same season of coming back to work that he shared this comment from chapter 4: "I work with a guy who is a disabled vet and another coworker said, 'So, I don't care. Why should I care? You all volunteered for it!'" The internal logic to a dominant ideology of "support the troops" (and its liturgical "thank you for your service") is that one way in which society extends its appreciation is by ensuring that the job the veteran put on hold in order to deploy will be waiting once they return. This logic, however, is not always this simple, and often conflicts with other dominant ideologies. Within the Reserve component, especially, when individuals step away from jobs to deploy, they expect a job when they return. There are, of course, legal mandates to protect veterans at their previous pay scale and job responsibilities, but what I am more interested in is a) how neoliberal reason creates subjects (both as employee and as employer) that value competition and productivity over solidarity even after putting their life on the line in combat, and b) how this neoliberal reason functions ideologically. Now, though, to make this neoliberal as subject argument, I need to briefly define neoliberalism, and more important for my argument, neoliberal reason.

Neoliberalism posits that deregulated markets are the key to personal freedom. Neoliberal theorists insist that governments not intervene in the market. Without intervention, capital is freely able to move across borders. With this in mind, neoliberal ideology wants to limit, or end entirely, state-sponsored social welfare programs. Neoliberalism's ideal is based upon the (erroneous) metaphor that a "rising tide lifts all boats." The implication being that if the government stays out of the market and enables common sense and deregulated trade to continue, then the trickle-down economics of President Ronald Reagan (aka "Reaganomics") will benefit the entire population. The question that remains, however, deals with those who have been systematically oppressed and prohibited from purchasing a boat, so to speak. The systematic critique reveals the underside of neoliberalism, and that this underside favors—and advocates for—competition, inequality over equality. Pastoral theologian Cedric Johnson asserts that neoliberalism "created space for an *acceptance of inequality* as an essential component of economic growth and social progress."[55]

What took place in Reaganomics is precisely what is programmed in neoliberalism: wealth is redistributed unevenly and benefits the minority who are already wealthy. The oppressed are further marginalized, with the gap between the wealthy and marginalized burgeoning. These groups are split on class lines. Pastoral theologian Bruce Rogers-Vaughn argues convincingly that class inequality is an additional element within the matrix of the neoliberal turn; inequality is a "form of oppression intrinsic to capitalism, in which dominant elites use their economic, political, and cultural power to subjugate and stigmatize people who do not possess such power."[56]

The lived experiences of the veterans interviewed lend credence to Rogers-Vaughn's insight. They each found themselves as one of the many working poor in this country. This, as detailed in chapter 4, was a reason for enlisting: financial stability to either purchase a home or receive tuition assistance. Without a college education, each veteran struggled to find steady work in which they were not underemployed, and this continued after the deployment as well. Each are reduced to neoliberal reason. Neoliberal reason puts forth that it is the economic realm, rather than a social or political, that determines existence. Critically, then, it becomes those realms of life that have *nothing* to do with economics that are subjected to an economized enterprise. As political theorist Wendy Brown offers, "the point is that neoliberal rationality disseminates the *mode of the market* to all domains and activities—even where money is not at issue—and configures human beings exclusively as market actors, always, only, and everywhere as *homo economicus*."[57]

At this point, multiple pastoral and practical theologians, such as Bruce Rogers-Vaughn, Cedric Johnson, James Poling, Ryan LaMothe, Nancy Ramsay, Philip Helsel, Stephen Pattison, Archie Smith, and Joyce Ann Mercer have offered liberative practices that could rectify or at least operate as resistance to veterans succumbing to neoliberal reason.[58] Undergirding their concerns is the critique that when everything is economized as capital, "the foundation vanishes for citizenship concerned with public things and the common good."[59] Employers, who are driven through competition to increase profits and productivity, have little room for military service in a Reserve component context. It is far easier to "support the troops" when those troops are on active duty and not a threat to one's financial bottom line. Once again: Lisa is perfectly on point when she says that she is not included in the narratives of combat; it is *always* some other veteran. Therefore, there is a conflict between dominant ideologies: one of supporting the troops and one of rampant atomized neoliberal capitalism. Of the names mentioned, I want to focus on Rogers-Vaughn as a clinical collaborator on the impacts of neoliberal reason.

In the midst of oppression caused by neoliberal reason, Rogers-Vaughn enacts practices to revitalize theology as an asset for care. In chapter 6, I offer counterhegemonic spiritual care and counseling practices, but at this juncture, Rogers-Vaughn provides a prescient critique for caregivers. The care provided can be one in which neoliberal reason is "promoted," "accommodated," or "resisted."[60] There are similarities between these three positions and Hall's three hypothetical ideological decoding of texts. Those care practices that *promote* neoliberal reason are those that maintain an individual pathologizing. The caregiver "urges us to look for the origins of suffering only within our selves."[61] Phillip's supervisor's comment above that reserve service members "volunteer for it" is a statement of neoliberal promotion. There is no liminal space in reintegration in which a veteran can rest—the market does not rest, and neither should you. This is the dominant-hegemonic reading of neoliberalism

Rogers-Vaughn argues that those who *accommodate* neoliberal reason attempt to put a reformed face on neoliberalism. These accommodating practices come up to the precipice of actual resistance and critique of the system, and similar to Hall's negotiated readings, back away at the crucial juncture. At this juncture, "the usual outcome, however, is that sufferers receive just enough help to remain conformed to a system that produced the pain to begin with."[62] An accommodated posture toward neoliberalism is precisely where the majority of moral injury interventions reside. The interventions are framed within the current ideological framework, and those frameworks are not challenged. They simply exist. Structural and systemic change does not happen within accommodation.

The goal is to confront the neoliberal and ideological systems and *resist* (Hall's oppositional reading). The systemic suffering that is imparted on one individual remains true for the community as well. This is the case because suffering (and inequality) is built *into* the system. The community is in this together, and this is precisely the solidarity that this section and the project overall advocate. Rogers-Vaughn adds, "individuals no longer suffer in isolation, nor are they alone responsible for 'getting better. . .' The solidarity it nurtures is caring in itself."[63]

IMPLICATIONS FOR PRACTICAL THEOLOGY AND THIS PROJECT

This chapter revealed and critiqued *how* society is mired within ideology and how ideology impacts returning veterans experiencing moral injury. The strategic advantage of utilizing Hall and Gramsci's work is that opposition is possible. Practices of opposition can be enacted. Those practices are the goal of chapter 6 as it will highlight the fourth task in Osmer's practical theology: the pragmatic. This task forms and enacts "strategies of action that influence events in ways that are desirable."[64] Osmer fundamentally understands the pragmatic as a leadership role. Leadership is a role of a Gramscian intellectual, which I argue is a role for military chaplains. Intellectuals are essential in privileging spaces in which new intellectual, ideological, and creative resources are disseminated. Remember, from chapter 2's genealogy on ideology, in Gramsci's understanding of hegemony there is never a universal political foundation, and intellectuals are instrumental in forming counterhegemonic communities that ultimately "permeate society with new systems of value, belief, and morality."[65] A counterhegemonic community sees veterans differently; a counterhegemonic community listens intently; a counterhegemonic community supports the lived experience of veterans. Therefore, it is now time to privilege counterhegemonic moral injury support groups, as, up until now, these practices of opposition have not been offered.

NOTES

1. Stuart Hall, "The Rediscovery of 'Ideology': Return of the Repressed in Media Studies," in *Culture, Society and the Media*, ed. Michael Gurevitch (New York: Methuen, 1982), 88.
2. Hall, "'Ideological Effect,'" 346.
3. Ibid., 340.
4. Ibid., 341.
5. Ibid., 342.
6. Ibid., 343.
7. Stuart Hall, "Encoding/Decoding," in *Culture, Media, Language: Working Papers in Cultural Studies, 1972–1979*, ed. Stuart Hall, Dorothy Hobson, Andrew Lowe, and Paul Willis (New York: Routledge, 1980), 136.
8. Ibid., 130.
9. Ibid., 136.
10. Ibid., 136.
11. Ibid.
12. Ibid., 137.
13. Ibid.
14. Ibid., 138.
15. Taken from the subtitle of Kyle's autobiography, Chris Kyle and Jim DeFelice, *American Sniper: The Autobiography of the Most Lethal Sniper in U.S. Military History* (New York: HarperCollins, 2012).

16. Ben Beaumont-Thomas, "Clint Eastwood: *American Sniper* and I are anti-war," *The Guardian*, March 17, 2015, accessed April 17, 2017, https://www.theguardian.com/film/2015/mar/17/clint-eastwood-american-sniper-anti-war.
17. Hall and Whannel, *The Popular Arts*, 36.
18. Kyle and DeFelice, *American Sniper*, 4.
19. A. O. Scott, "Review: 'American Sniper,' a Clint Eastwood Film With Bradley Cooper," *New York Times*, December 24, 2014, accessed April 17, 2017 https://www.nytimes.com/2014/12/25/movies/american-sniper-a-clint-eastwood-film-starring-bradley-cooper.html?referrer=google_kp&_r=3.
20. Michel Foucault, "Film and Popular Memory" in *Foucault Live: Collected Interviews, 1961–1984*, ed. Slyvère Lotringer, trans. Lysa Hochroth and John Johnston (New York: Semiotext(e), 1996), 125.
21. Kyle and DeFelice, *American Sniper*, 3.
22. Seth Rogen, Twitter post, January 18, 2015, 11:05 a.m., https://twitter.com/sethrogen/status/556890149674434560?lang=en.
23. Michael Moore, Twitter post, January 18, 2015, 3:40 p.m., https://twitter.com/mmflint/status/557250871386718209.
24. Newt Gingrich, Twitter post, January 18, 2015, 5:37 p.m., https://twitter.com/newtgingrich/status/556988852897054720?lang=en.
25. Hall, "Encoding/Decoding," 136.
26. Ryan LaMothe, "Men, Warriorism, and Mourning: The Development of Unconventional Warriors," *Pastoral Psychology* 66, no. 6 (December 1, 2017): 820.
27. Clifford Green, *Bonhoeffer: A Theology of Sociality*, rev. ed. (Grand Rapids, MI: William B. Eerdmans Publishing Company 1999), 20.
28. The Marxist underpinnings are arguably more pronounced within Latin American liberation theology. Cornel West has noted the perplexing nature of the relationship of "strangers" between first-generation Black liberation theologies (specifically James Cone and J. Deotis Roberts) and Marxism, in "Black Theology and Marxist Thought," in *Black Theology: A Documentary History, Volume 1: 1966–1979*, ed. James Cone and Gayraud Wilmore (New York: Orbis Books, 1993), 409. West offers a Marxist class and economic analysis that overlaps within Black liberation theology. The overlap includes both systems emphasizing the plight of the oppressed, their powerlessness, and the possibility of empowerment.
29. Bonhoeffer, *Letters and Papers*, 17.
30. Segundo, *Liberation of Theology*, 32.
31. Green, *Theology of Sociality*, 3, 43n55.
32. Dietrich Bonhoeffer, *Ethics*, Dietrich Bonhoeffer Works, vol. 6, ed. Clifford J. Green, trans. Reinhard Krauss, Charles C. West, and Donald W. Stott (Minneapolis: Fortress Press, 2009), 267. Emphasis in original.
33. Ibid., 284.
34. John A. Phillips, *Christ for Us in the Theology of Dietrich Bonhoeffer* (New York: Harper & Row, 1967), 28.
35. Bonhoeffer, *Sanctorum Communio*, 21.
36. Ibid., 141.

37. Eberhard Bethge, *Dietrich Bonhoeffer: Theologian, Christian, Contemporary*, trans. Eric Mosbacher, Peter Ross, Betty Ross, and Frank Clarke (London: William Collins Sons & Co., 1970), 4.
38. Ibid., 22.
39. Christiane Tietz, *Theologian of Resistance: The Life and Thought of Dietrich Bonhoeffer*, trans. Victoria J. Barnett (Minneapolis: Fortress Press, 2016), 10.
40. Bonhoeffer, *Sanctorum Communio*, 34.
41. Bonhoeffer's view is contrasted most strikingly with Martin Buber's "I-Thou." Buber's classic *I and Thou* (New York: Scribner Classics, 2000) was published in 1923, four year before the publication of the dissertation version of Bonhoeffer's *Sanctorum Communio*. Buber's I-Thou was preferred over and against an I-It relationship. Buber's formula is more intimately relational (overcoming an objectified "It" mentality), while Bonhoeffer's I-You develops an ethical boundary, or limit, between I and You. Buber is not cited in Bonhoeffer's bibliography.
42. Bonhoeffer, *Sanctorum Communio*, 51. Emphasis in original.
43. Ibid., 54–55. Emphasis in original.
44. Ibid., 88.
45. Ibid., 80.
46. Ibid., 145. Emphasis in original.
47. Ibid., 178. Emphasis in original.
48. John Patton, *Pastoral Care in Context: An Introduction to Pastoral Care* (Louisville: Westminster John Knox Press, 1993), 6.
49. Ibid., 39–61.
50. Stated clearly, as Kathleen Greider, Gloria Johnson, and Kristen Leslie note in their exhaustive look at feminist writings within pastoral theology, feminist writings have "contributed precisely and significantly to the emergence of this communal-contextual paradigm." Kathleen Greider, Gloria Johnson, and Kristen Leslie, "Three Decades of Women Writing for our Lives," in *Feminist and Womanist Pastoral Theology*, ed. Bonnie J. Miller-McLemore and Brita L. Gill-Austern (Nashville: Abingdon Press, 1999), 22.
51. Carroll A. Watkins Ali, *Survival and Liberation: Pastoral Theology in African American Context* (St. Louis: Chalice Press, 1999).
52. Ibid., 9.
53. Ibid., 121. This is done to create a more robust version of care, not eradicate Hiltner's work.
54. Hall, "The Neoliberal Revolution," 334. Emphasis in original.
55. Cedric C. Johnson, *Race, Religion, and Resilience in the Neoliberal Age* (New York: Palgrave Macmillan, 2016), 37. Emphasis in original.
56. Bruce Rogers-Vaughn, "Class Power and Human Suffering: Resisting the Idolatry of the Market in Pastoral Theology and Care," in *Pastoral Theology and Pastoral Care: Critical Trajectories in Theory and Practice*, ed. Nancy J. Ramsay (Malden, MA: Wiley Blackwell, 2018), 55.
57. Wendy Brown, *Undoing the Demos: Neoliberalism's Stealth Revolution* (New York: Zone Books, 2017), 31. Emphasis in original.

58. Rogers-Vaughn notes that the locatedness is an impactful aspect of Johnson's work. Johnson is able to "explore the impact of neoliberalism on African Americans," *Caring for Souls in a Neoliberal Age* (New York: Palgrave Macmillan, 2016), 23. This is an important difference between Johnson and Rogers-Vaughn's monographs. Rogers-Vaughn methodically traces neoliberalism's development from a broader perspective and focus on the "global community" in *Caring for Souls in a Neoliberal Age*. Johnson, on the other hand, reverses Rogers-Vaughn's move: Johnson takes a similarly methodical analysis and explores the material effects of neoliberal on a specific African American community in New York City. Helsel is interested in power as the manifestation of class struggle. LaMothe's work focuses on how neoliberalism neglects care practices. Stephen Pattison helpfully, and successfully, integrates specifically Latin American liberation theologies with pastoral care functions in a British psychiatric hospital setting. Mercer mentions the absence of class in an analysis of oppression, but merely addresses the absence and does not offer substantive additions or enact practices.
59. Brown, *Undoing the Demos*, 39.
60. Rogers-Vaughn, "Class Power and Human Suffering," 71.
61. Ibid.
62. Ibid.
63. Ibid., 72.
64. Osmer, *Practical Theology*, 176.
65. Emilie M. Townes, *Womanist Ethics and the Cultural Production of Evil* (New York: Palgrave Macmillan, 2006), 21.

Chapter 6

Oppositional Forces

Toward a Counterhegemonic Paradigm for Spiritual Care and Counseling

The goal of this final chapter is to propose an interdisciplinary practical theology that is constructive, critical, and offers transferable proposals for veteran support for military chaplains and spiritual care providers.[1] In light of my participants' MIEs and experience of reintegration—highlighted and analyzed in chapter 4—and the communal-contextual commitments established in chapter 5, I now want to discuss how reintegration could function differently in the midst of American dominant ideologies. In recognition of the demonstrated power of how dominant ideologies can amplify existing moral injuries, one in which reintegration is further complicated by a cultural ecosystem steeped in ideology, the framework that I am developing is a modified spiritual care and counseling paradigm based within Lartey's "social action," in which the tasks of resistance, opposition, and solidarity function to "speak truth to power."[2] Each of these functions are communal; reintegration is a communal activity for both the veteran and the civilian community.

The framework I propose is an interdisciplinary practical theology: first, from cultural studies, the entire framework is understood within the struggle of counterhegemony offered by Gramsci and Hall. Within that framework, military chaplains are positioned to provide effective leadership against an exacerbation of moral injury through dominant ideologies. The positioning of military chaplains is as Gramscian "intellectuals," originally outlined in chapter 2. Upholding that framework, from Christian theology, is a God that is co-suffering with the veteran. Finally, I develop a liturgy of moral injury support that emphasizes that co-suffering God. With that, I return to Hall, and his development of Gramsci's political theory.

Hall proposed hypothetical interpretive positions for decoding ideology. Practically, this means that the work of reintegration and the specific praxiological commitments are context dependent. Resistance to dominant ideologies is possible. I am, therefore, utilizing Hall's oppositional readings with the potential of forming a counterhegemonic community from a Gramscian perspective. Utilizing Gramsci, in particular, is necessary as he offers active, concrete, and tangible steps toward change. Gramsci, then, is beneficial in developing a framework capable of combining the complexities of moral injury reintegration, the matrices of dominant ideologies that exacerbate reintegration, and the role of military chaplains to provide support.

In my own interlocution of Gramsci, I am not "simplistically believing Gramsci has the answers or holds the key to different historical and contemporary problems," but I am resolute that there is an "importance of thinking in a Gramscian way."[3] What this project is attempting to critique are dominant ideologies concerning US military service and how these ideologies function to exacerbate an existing moral injury. A dominant ideological narrative of US military service, as Moon reminds us, "honors military service on a superficial level and cannot easily accommodate evidence of PTSD, moral injury, and veteran suicide because these phenomena seem to diminish the stoic warrior image."[4]

A dominant ideology and mythology of the US military (and service members) is the belief in a quasi-holy cause: if the United States is involved in a conflict, there is an inherent threat to US security. The logic suggests it is better to fight "them" over "there" lest "they" bring the conflict to the United States. The service members are the ways in which security is maintained. As conduits of our national security, they remain removed from critique, and deserve our platitudes and valorization. Dominant ideologies are produced, as I have shown in numerous places, in our reverence to the service member via the U.S. flag during the national anthem at sporting events, our liturgical mantra of "thank you for your service" disconnected from actual appreciation, and within mass-media texts (promulgated in particular within *American Sniper*). The veteran is the centerpiece of this dominant ideology.

Dominant ideologies mold the cultural consciousness of veterans and civilians alike. What is needed, and what I have been building toward, is an oppositional resistance that pushes back against ideological domination. Lisa, Andrew, Angela, and Phillip—like many veterans—noted that they needed a community that functioned differently; I propose that a counterhegemonic community functions in just that way.

I ended chapter 5 noting that within an oppositional reading and the counterhegemony it offers in resistance, it becomes necessary to dislodge the media reaction, and instead focus on realities immersed in a veteran's lived experience. To elaborate, above it has been critical and necessary to focus on *theoretically* what is taking place in dominant ideology's production and how dominant-hegemonic readings are produced within professional ideology.

Practical theology is able to speak back into Hall's work and expand upon his oppositional readings through the inclusion of spiritual practices that cultivate lives *of* opposition. An oppositional reading seeks to retotalize reintegration through addressing the lived experience and context of the veterans and civilians involved in the counterhegemonic community. My own practical theological interpretation of Hall's oppositional reading position is strengthened through Hall's Gramscianism. I will argue that military chaplains primarily provide solidarity, as religious leaders and Gramscian intellectuals. However, before making an argument about the *posture* of military chaplains, I need to "think" from a Gramscian perspective how a counterhegemonic force strategically helps veterans reintegrate.

THE CHAPLAIN AS GRAMSCIAN INTELLECTUAL

In the previous chapter's examination of Stuart Hall I discussed his critique of Althusser that you cannot formulate *a single* ideology for an entire class. Media texts are produced—not given—within dominant ideologies, and I noted Hall's positioning concerning opportunities for "oppositional" readings of texts. An oppositional position "detotalizes the message in the preferred code in order to retotalize the message within some alternative framework of reference."[5] Veteran support is in desperate need of both a retotalized understanding of reintegration and also an alternative framework for reintegration. Throughout this chapter, the role of the chaplain is fundamental to the paradigm. This is partly because of the chaplain corps' place within the military culture, the aforementioned ministry of presence, and therefore since chaplains are within that culture—even on the periphery—there is still a unique opportunity to journey with veterans coming to terms with MIEs. Before discussing a proposed liturgy and mechanisms of solidarity, I want to return to Gramsci as a means of solidifying the role of chaplains.

A "War of Position": Gramsci's Counterhegemony

Oppositional readings are the conduit within which a Gramscian counterhegemonic "cell" can take shape. Within Gramsci's seventh prison notebook, dated between 1930 and 1931, a counterhegemonic group can take power via the continuum of political struggle between a "war of maneuver" and a "war of position."[6] These two metaphors, and they typically *remain* metaphors for Gramsci, have their own complex lineage within the Marxist tradition. Of note, Engels utilized the phrase (based on his reading of Napoleon), and both Vladimir Lenin and Leon Trotsky (adapting from Carl von Clausewitz) also deploy the phrase.[7] I am resolute that through my usage of Gramsci's modification of the phrase I can imagine the potentiality of fresh nonviolent tactics for counterhegemonic proposals.

The continuum was how Gramsci conceptualized the differences in revolutionary potential between Western Europe and Russia. In Russia, a "war of maneuver" entailed the physical movement of a working-class element, small in number, to take power back. A small group, split on class lines, could overwhelm the state because the level of civil support offered no resistance. Critically, then, Gramsci's point is that a state's hegemony is stabilized through its power of coercion within civil society *and* by its use of force (the Machiavellian centaur mentioned in chapter 2). Direct action fails when the state's credibility is firmly ensconced in civil society. By contrast, in Western Europe, civil society represents a robust interplay of hegemony.

Therefore, that same small faction that ignited revolution in Russia would not work within the Western European model due to the established civil society apparatuses in which intellectuals, the "organizers of ideology," spread common sense values and norms to the masses.[8] Gramsci elaborates, "in Russia, the State was everything, civil society was primordial and gelatinous; in the West, there was a proper relation between State and civil society, and when the state trembled a sturdy structure of civil society was at once revealed."[9] A war of maneuver, then, would ultimately fail in Western Europe. By contrast, a war of position is the strategic means of resisting domination.

Gramsci's war of position is the activation and building of a counterhegemonic force. To build this cell, Gramsci noted the necessity of leadership and intellectual resources not found within the hegemonic institutions. This point is elaborated in a lengthy—yet critical—section of the *Quaderni*:

> A human mass does not "distinguish" itself, does not become independent in its own right without, in the widest sense, organizing itself; and there is no organization without intellectuals, that is without organizers and leaders, in other words, without the theoretical aspect of the theory-praxis nexus being distinguished concretely by the existence of a group of "specialized" in conceptual and philosophical elaboration of ideas. But the process of creating intellectuals is long, difficult, full of contradictions, advances and retreats, dispersals and regroupings.
>
> The process of development is tied to a dialectic between the intellectuals and the masses. . . . But every leap forward towards a new breadth and complexity of the intellectual stratum is tied to an analogous movement on the part of the "simple."[10]

Hegemony is spread and further established in a society through the success of intellectuals and their implementation of consent. Victory is the "seizing of the balance of power . . . commanding the balance of political, social, and ideological forces at each point in the social formation."[11] Intellectuals are arguably the second most important theme in Gramsci's *Quaderni*, behind hegemony, and were drafted in his journals mere months following his arrest.

For Gramsci, an organic intellectual is responsible for organizing and leading from within their class structure. By "organic," Gramsci has in mind "a revolutionary urgency, the look forward to winning the struggle for hegemony."[12] Organic intellectuals ensure that new ideas filter to the masses. These ideas are not forced upon people in a hierarchical, top-down fashion as propaganda; rather, they seep into everyday life as values, language, and culture. Organic intellectuals represent the aspirations of their own class, and although this project and veteran support are not solely fixated on class, military chaplains have class interests and alliances because they represent the service members in the unit and are bonded to veteran communities via the solidarity of shared experiences. With that background, a more thorough analysis of organic intellectuals is warranted, and specifically how military chaplains can function as Gramscian intellectuals.

Military Chaplains as Gramscian Intellectuals

Military chaplains can function as Gramscian intellectuals, and critically for this project, military chaplains *as religious leaders* and intellectuals can stand in as a political "party." Gramsci elaborates this as, "it turns out that on special occasions the clergy of all the churches has functioned as public opinion in the absence of a normal party and a press organ of such a party."[13] *Prison Notebooks* editor and Gramsci scholar Joseph Buttigieg offers some clarity on how clergy functioned in this way. Gramsci was particularly struck by an active clergy role in political strikes, campaigning for an eight-hour workday in the 1910s and 1920s, and allowing parishioners to use church property for strikes. Gramsci noted individual ministers partaking in these strikes, but also the ecumenical commitments of the Industrial Committee of the Protestant Churches, the Federal Council of Churches, and the Interchurch World Movement.[14]

Now, clearly, I am not advocating the advancement of a new political party. Within the *Quaderni*, "party" refers to a traditional political party, and to a "wide range of organizations that bring those with common interests together."[15] To make this point even more clearly, Hall describes the importance of parties in Gramsci: "no ideology or theory is worth its salt until it has found a party, that is to say, an organizational-institutional expression."[16] Therefore, with Gramsci's insights, it is clearer how chaplains can function in an intellectual role supporting counterhegemonic cells and ultimately empowering the veterans toward their own liberation.

The churches, then, represented, in lieu of a political party, a cell that single-mindedly represented the interests of that cell. Within that paradigm, clergy standing in for political interests of a cell, Gramsci begins to sketch what becomes an organic intellectual. The role of clergy is an essential component not only of how Gramsci understands intellectuals but also how he methodologically conceptualizes the alliance of intellectuals and the working class within an implementation of communism. This ultimately becomes a historical search of clerical functions for Gramsci.

Beginning in the Middle Ages, Gramsci introduces the "class" role of the clergy. He notes that a study of the class role of the clergy:

> It seems to me that it would be indispensable as a beginning and as a condition for the whole study that remains to be done on the function of religion in the historical and intellectual development of humanity. The precise juridical and de facto situation of the Church and the clergy in various periods and countries, its economic conditions and functions, its exact relations with the ruling classes and with the state.[17]

Gramsci's insights are important to my overall investigation: how can a military chaplain, within a hegemonic institution of the military-industrial complex, still maintain the interests of their service members? In another place in the *Quaderni*, Gramsci notes that, even within hegemonic institutions, some factions and cells can work against hegemonic tendencies. Gramsci notes that "modernists, integralists, and Jesuits" are the cells within the Roman Catholic Church fighting against its hegemony. Those groups are "'parties' inside the 'international absolute empire.'"[18]

I want to return to the Middle Ages, as Gramsci notes that the clergy operated as their own class, yet still operate with a dialectical tension in regard to the interests of feudal parishioners. On one hand, clergy positioned themselves in alliance with the peasants "against the other classes, insofar as the peasants enabled the Roman Catholic Church to maintain and expand its influence."[19] There is, therefore, a religious and moral commitment to the lower classes and the clergy's own economic interests. However, Gramsci highlights how the Roman Catholic Church exploited their parishioners. Clergy still aligned themselves with the noble class continuing to support their own economic interests. This commitment to both sides continues into the French Revolution. Gramsci notes that during this period, the clergy's corruption fractured itself as a class and also fractured clergy from their parishioners.

Moving into capitalism, Gramsci starts to refer to clergy as a "caste." Clergy—and the church—have lost their cultural hegemony and prominence. Clergy sought to maintain relevance through education. The Roman Catholic

Church begins to teach—Gramsci is most mesmerized by the University of the Sacred Heart (University of Sacro Cuore)—activities to become "the mechanism for selecting the most intelligent and capable individuals from the lower classes to be admitted into the ruling class."[20] The mechanisms are not merely reproducing clergy; on the contrary, these mechanisms work to place lay individuals into key leadership roles in society. These lay individuals are "more valuable auxiliaries of the Church as university professors rather than as cardinals, etc."[21]

Ultimately, Gramsci's interest in the clergy is as a paradigm and fermenting link between the working class and the intellectual in his time. He is interested in how the clergy align themselves, via class, with their parishioners. Gramsci's interest decisively returns to my focus on the role of the military chaplain *as an intellectual* within the military ecosystem.

Military chaplains are educated professionals. Through attaining a bachelor's degree and a master's degree, military chaplains represent some of the more educated professionals in the military—even if their graduate education is theologically based, or as Gramsci describes, "intellectually subaltern."[22] That education affords military chaplains a commissioning status as officers in the military. In a rank-based ecosystem of power, chaplains occupy a space of privilege that comes from rank, including direct access to the commanding officer. How then should chaplains use power and privilege? Chaplains must make a conscious decision about how one uses the power and privilege afforded to them by the system.

Gramsci's intellectuals are a necessary component of the struggle toward power, and this could potentially present a conflict for military chaplains. For example, chaplains, as officers, are compensated at a higher rate than lower-rank enlisted soldiers. This reality does not remove the necessity to remain in solidarity and aligned with their soldiers. Gramsci notes that this potential conflict is not reserved for modern-day clergy. Clergy have always had the potential to function in a liminal space, on the one hand, representing their own economic interests while, on the other hand, also supporting the religious interests of their parishioners.

Returning to the experience of moral injury and the role of military chaplains, class does not restrict the experience of moral injury. Chaplains, and high-ranking officers, are impacted and traumatized just as lower-ranked enlisted soldiers—which can further solidify a chaplain's link to the lived experience of the soldiers reintegrating moral injury. Military chaplains, as uniformed members of military service, are immersed within military life and military culture. Military chaplains serve in space and time with the individuals struggling to reintegrate, and this is an asset for providing care. It is further the precarious nature of the military chaplain within the military that affords them the ability to strive for new possibilities. These are precisely Gramscian concerns; namely, the chaplain is linked to the concerns of the service members. Political theorist, theologian, and Marxist Roland Boer states:

> Gramsci is intrigued as to how one's own "caste" may operate at some remove from the current structures, always with an eye on a very different future. Thus, precisely through being a relic, the clergy provide a glimpse of something different, able to look forward in a way that does not merely replicate the present. It is their backwardness, the fact that they are not in touch that enables them to anticipate a different future. That is to say, they act according to an agenda that is, in many respects, their own and not "of this world."[23]

It is the precarious liminality of chaplains that enables them to strive toward healing.

Pastoral and practical theologian Stephen Pattison offers his own liberative *leitmotif* to this liminality, *locus theologicus*, as a reminder that spiritual care and counseling is done from a location embedded in a community.[24] Pattison helpfully articulates a spiritual care and counseling *posture*; namely, one that is generated from located solidarity. Womanist theologian Carroll Watkins Ali argues similarly that power and privilege must be attached to a conscious solidarity or an "attitude of advocacy" for the "least of these who are struggling with survival and liberation issues."[25] Chaplains are responsible to provide support to service members assigned to them, and I will expand this to note that even during the reintegration season when veterans have completed a deployment, these individuals are still within a chaplain's *locus theologicus*. Chaplains understand the plight of these veterans and are therefore situated to provide clinically competent support. For example, although I did not deploy with the veterans in this project, we have solidarity. I find resonance in that I am transformed through their experiences—as they intersect with my own—and work to provide solidarity with them in these contexts. I want them to flourish now that they are home.

To conclude this section on Gramsci and intellectuals, it is clearer now that military chaplains can function in an intellectual role. In the post-9/11 landscape, military chaplains are understood as "force multipliers." This role functions to elevate a chaplain's place within military tactics—essentially, how can a chaplain get service members back "in the fight" more effectively? Force multiplier is a divergence from a traditional role of merely providing religious support to assigned service members. Within the function of offering religious support, a chaplain could provide the religious rites and services he or she is authorized to perform as an ordained or credentialed religious leader. However, at present, military chaplains are tasked with advising the command on issues of morale, morals, ethics, and religion and *how* these factors impact the mission. In combat, especially, the military expects chaplains to rally behind the cause of the country. Theologian Ed Waggoner, drawing from Pentagon documents, states that chaplains "as a multiplier of force . . . speak martially, patriotically, and divinely all at once."[26]

At face value, this is absolutely a chaplain's role. It is essentially asking the chaplain to report on the morale of the service members. However, Waggoner shows how this role of "moral calibration" has the potential of being cloaked within a chaplain's "soft-power."[27] A chaplain is tasked to counsel service members on myriad issues that sprout from military service. This is a good thing. Chaplains, as the only personnel in the military ecosystem who hold complete confidentiality of privileged communications, can grasp the psychospiritual issues that are prevalent in combat. However, according to Waggoner, "The military wants chaplains to dispel prejudice against the enemy, but forbids them to promote an exchange of cultural or religious perspectives successful enough to 'mitigate operational requirements and use of military force if necessary.'"[28]

Waggoner's concerns are valid: chaplains need to know, when they are asked about the morale of service members, is the question coming from a place of assessing combat strength or from a genuine concern for the men and women? I appreciate how complex that line can be. The role of a force multiplier is a complex position; yet, the chaplain's role as an intellectual is more conducive to the long-term health of veterans. A force multiplier has little concern for long-term implications of getting a service member back "into the fight," who may be suffering psychospiritually. All this to say, hastening a service member prematurely back "into the fight" can generate new issues—or exacerbate existing concerns, which is why I advocate an additional role of an intellectual. An intellectual role, then, provides service members with the mutual empowerment to move toward healing. Returning to the overall movement of this chapter, I want to revise practices through the insights of this case study. Pattison's *locus theologicus* takes the chaplain only part of the way. Concrete practices are necessary to go the distance.

LIFE TOGETHER: COUNTERHEGEMONIC COMMUNITIES

In chapter 4, Phillip lamented that upon reintegration from combat his prayer life was not the same, and his relationship to his previous religious community was not the same. He changed after combat, and needed a religious community that could adapt alongside him. Lisa lamented that she needed a community that could sit in the tension—not necessarily "fix" her, but journey alongside her. Religious communities should assist veterans in locating or cocreating a reintegration-ready theology; a theology for reintegration, and one that stands in opposition to easy answers about war, society, and theology. I now want to return more specifically to how reintegration could function within Hall's "retotalization." Hall's usage of retotalization is a helpful reminder in our thinking about reintegration: our current mode of bringing veterans home is not working.

A dominant-hegemonic ideology that insists that our societal refrain of "thank you for your service" and the various participating restaurants on Veterans Day and Memorial Day represent a sufficient salve to our war effort over the past twenty years is isolating, egregiously dangerous, and, importantly, not cajoling our country any closer to *concluding* these wars. Andrew knew that the phrase was a token, perhaps a token born out of the shrouded mystery of military service. Former army chaplain and founder of the Austin, Texas, chapter of the Episcopal Veterans Fellowship David Peters states the matter succinctly: "Veterans do not need more barbeques, picnics, or trips to amusement parks. These are all nice and I have enjoyed most of them. What we do need is community, connection to ourselves, each other, and to God. In my view, the Church is the best organization to do this."[29] Andrew commented that something more like "thank you for your sacrifice" speaks to the reality of military service. All service members sacrifice, regardless of combat experience. Sacrifice encompasses everything that envelopes military service, and sacrifice diminishes the empty gesture of service.

Therefore, Hall's oppositional reading position provides the creativity to ask different questions and live different lives. Living different lives in the midst of a dominant American ideology is *not* easy. A responsible caregiver needs to claim this reality from the beginning. To stand up for our veterans—to truly take a stand—requires we "renounce the privilege of ignorance that the present-day American wars and to the extent that they renounce generalizations—promilitary, antimilitary, pro-US-foreign-policy, anti-US-foreign-policy—in favor of close and sometimes painful attention to the war-torn bodies among them."[30]

The retotalization starts within religious communities, as it is religious communities that are situated to handle the spiritual distress, the MIEs, the ethical concerns of war's moral ambiguities, and assist in re-authoring and externalizing previously unhelpful spiritual resources for some veterans. Finally, religious communities and their leaders, can function as "benevolent moral authorities."[31] Military chaplains, as ones who can embody benevolent moral authorities, are situated to enact retotalized spiritual practices that assist in the reintegration of the veteran back into the religious community.

Healing takes place *within* communities in which both sides have something at stake; Phillip and Andrew sought a reintegration that included neither unwarranted praise nor ostracizing, and they are able to acknowledge that some civilians are looking for meaningful ways to show gratitude for his sacrifices. The difference, for Phillip, was while deployed, individuals supported him. They were in close proximity. His "battle buddies" knew Phillip and knew when he was struggling. Significantly, they knew Phillip's story. When Phillip came home, and attempted to reintegrate his experience, he was alone—he was not surrounded by people who knew his story. Lisa spoke to this reality too. Combat, even in the kinetic and highly chaotic environment, has a seductive simplicity to it. However, her experience as well as Angela's, include the harrowing morally injurious events of ongoing sexual assaults and harassment. Back home, then, the intimacy of combat subsided into loneliness. A goal of what follows is to elucidate that the same communal intimacy previously available in combat is still available during reintegration. Therefore, how might counterhegemonic communities model religious practices of what I am calling a Liturgy of Solidarity? Counterhegemonic communities can "oppositionally" offer support to those suffering from an ideologically amplified moral injury.

Within reintegration and moral injury support, an "oppositional" reading provides two possibilities: for the veteran, it generates a process of normalization. This normalization concerns the veteran's humanity, not necessarily the normalization of the MIE. The veteran will come to discover that the MIE fits as a smaller episode within the entire narrative of the toll of these wars. Second, for the civilian community, a posture of deep solidarity through listening, understanding, and support affects how one looks at the world. After learning the stories of MIEs and the impact they have on lives, the lived experience *should* impact how one thinks about veterans' issues, spends money, or votes. It should also affect how we think about foreign policy decisions overall. We should become concerned with how our decisions impact communities in other countries as well. The counterhegemonic community provides the civilian with the space to reflect on how these conflicts take place in their name as well.

To elaborate further on this point, some civilians do want to help; there is helplessness, though, of being caught up in current ideological malaise. The difficulty has been creating spaces in which civilians *and* veterans attend. The civilian and military divide is real, and two decades of war have only widened that division. There *is* gratitude. There *is* support. What is missing, however, is the next step: specific and proactive steps of support to truly model gratitude. Counterhegemonic communities provide the space for civilians to be present to these stories and bear the emotions they contain.

The schematics of a retotalized liturgy of solidarity will vary depending on the veteran and their level of trauma; however, with our participant veterans' narratives in mind, there are certain aspects of their stories that are transferable to others experiencing the amplifying effects of moral injury. The liturgy contains spiritual practices that remain relatively ordered and unchanged throughout the sessions. The established schematics leave room for spontaneity as well. The spontaneity is present within a veteran's telling and retelling of the MIE. Elaine Ramshaw, in her formative text *Ritual and Pastoral Care*, describes the tension between order and spontaneity as:

> The need for order and continuity is fundamental to the ritual purpose. When people say, "This is the way we have always done it," they are saying something very precious in our fast-changing, mobile society. . . . This does not mean, of course, that nothing can never be changed. It does mean that, even more than in other areas of congregational life, the introduction of change in ritual practice must be gradual and respectful of the need for continuity of practice.[32]

The ritual that follows is adapted from my own Christian denominational context, the United Church of Christ. The intent of this adaptation is simple: an established and consistent ritual provides continuity and predictability for the community. Further, as Ramshaw notes, "the pastor's role is to assist in the people's creative task, through her knowledge of the church's liturgical tradition and the people's ritual needs."[33]

The schematics of this liturgy will also vary depending on civilian participation as well. This variance is not insurmountable; rather, it takes proactivity on behalf of the counterhegemonic community to include civilians in this reintegration liturgy of solidarity—civilians need reintegration as, like it or not, we have all been at war. The counterhegemonic community needs intentional spaces for co-constructing counternarratives of the war, MIEs, and the reintegration experience.

There is one vital caveat before unpacking the liturgy, and that is the overarching need for—and establishing of—safety. Before a veteran is invited to share a story that perhaps he or she has not shared with anyone else there must be, within that community, a commitment to creating a sacred covenant between veterans and civilian participants. The implementation of some basic covenantal assurances of confidentiality ensures a safe(r) environment for all involved. Besides the necessity to honor and maintain confidentiality, a commitment to deep listening is essential. This listening is not merely the lack of talking; rather, deep listening becomes a posture of solidarity as well. To listen, without the intent to solve, is a gift that we provide to our fellow counterhegemonic sojourners.

A LITURGY OF SOLIDARITY

The chaplain, as a Gramscian organic intellectual and facilitator of the counterhegemonic group, models the response to one's traumatic anguish: one in which we accompany, in solidarity, our fellow persons in their memories and pains. The Gramscian intellectual fuses and imbues language into a culture becoming common sense, and that most certainly applies to this liturgy. Using Phillip's narrative of religious resources, a Liturgy of Solidarity would need to include *at minimum* an intentional space for a reexamined prayer life and a religious community that privileges the telling of *a* story. Further, Lisa was more intentional about what she needed. She noted that a necessity of time and space brought her to some peaceful healing. This came through individuals who affirmed her and accepted her, regardless of the stories she shared. What this comes down to though, on the civilian side, is an ability to *share* in the trauma. We must all come to accept and realize the ways in which we perpetuate ongoing wars. Lisa noted, "don't thank me, help me." Beyond these practices, a dedicated space for absolution is vital to the overall reconciliatory movement of this moral injury reintegration liturgy.

Greetings

To begin, the chaplain or designated facilitator welcomes the community together. In this signaling the facilitator seeks to orient participants to the space. After going over the covenantal assurances, the facilitator recites a modified version of Psalm 51:

> *Facilitator*:
> Have mercy on us, O God,
> we unconventional warriors,
> according to your steadfast love;
> in your great compassion make us mindful of our offenses.
> Those offenses imparted,
> and those offenses we bear.
> Wash us through and through,
> from our complicities.
> May you surround us
> and sustain us
> when our transgressions encircle us. *Amen.*

Following this adaptation of Psalm 51, the facilitator leads the community in prayer:

> *Facilitator*:
> Holy One,
> As we come, again, together before you and our community,
> remind us of those memories we bring,
> the memories too burdensome to bear,

> and remind us that You carry them too.
> May each individual in this community feel your presence,
> and be surrounded by your grace and peace.
> May this community be a community of neighbors,
> journeying together. *Amen.*

The liturgy starts with this greeting and prayer to not only signal that our time is beginning, but further to proclaim the power of memory in liturgy. The prayer signals what it is that this community is about. In this space and in this time, you are not forgotten. In a liberative liturgy, it is of critical importance that the participants realize that a liturgy of solidarity is not committed to maintaining the current societal reintegration work. The practices within this liturgy seek to be vulnerable, challenging, and restorative.

Following the greeting, the facilitator or a designated member of the community begins a candle-lighting litany. The purpose of this practice is to further signal the intentionality of the space. The facilitator first lights a red candle, to symbolize the bloodshed from war. This bloodshed can symbolize the MIE that has brought the participant to the community, the death of loved ones during combat deployments, or the bloodshed of countless civilians in the countries where the U.S. is currently (or formerly) deployed. Further, red symbolizes—within the United Church of Christ—certain holidays, such as Memorial Day. Also, importantly, red represents the blood of Jesus Christ on the cross; therefore, this candle signals our mourning and complicities in our transgressions. Red, finally, also symbolizes the giving of the Holy Spirit at Pentecost, which reminds our participants of God's presence with us in our suffering. Following the lighting, this prayer is offered:

> We light this candle as a reminder of our pain, loss, and mourning;
> may this candle illumine in our hearts a reminder of all those who have suffered,
> in the midst of war.
> We mourn those on both sides—as there are *no* sides;
> just Your created humanity.
> Lead us toward healing. *Amen.*

The final candle lit in the litany is a green candle. Green, within the UCC, represents growth. Green is, therefore, a reminder that as a community we are moving forward together. The counterhegemonic movement is *always* an emergent struggle. Following the lighting, this prayer is offered:

> We light this candle as a reminder of our rejuvenation and growth;
> may this candle illumine in our hearts a reminder of our struggle for peace,

in the midst of war.
We press on, together, to restore, reconcile,
and yearn for peace.
Lead us toward healing. *Amen.*

Following this litany, the liturgy moves into a time for sharing our stories.

Sharing Our Stories

The liturgy of solidarity provides the space to tell *a* war story, not *the* war story. As I reflect on my own reintegration, I have yet to tell *the* story of my combat experience. Certain moods, friend cohorts, or memories elicit different narratives from my season in Afghanistan. Tim O'Brien comments on how time, memory, and war stories co-relate: "Often in a true war story there is not even a point, or else the point doesn't hit you until twenty years later, in your sleep, and wake up and shake your wife and start telling the story to her, except when you get to the end you've forgotten the point again."[34]

The telling of a war story may be more difficult for civilians to fully understand. War stories are not like other stories. O'Brien's classic *The Things They Carried* explicates this reality exquisitely. I would recommend each counterhegemonic community read *The Things They Carried*, and in particular the chapter, "How to Tell a True War Story." Few works of historical fiction get the reintegration experience as spot-on as O'Brien. Each member of the counterhegemonic community must remember "a true war story is never moral. It does not instruct, nor encourage virtue . . . if there's a moral at all, it's like the thread that makes the cloth. You can't tease it out. You can't extract meaning without unraveling the deeper meaning."[35]

The telling of a war story is not always a verbal affair. With Phillip's penchant for music—and specifically the ways in which music brought him back from war—art is a vital space to process and tell the stories of war. In a helpful February 2015 *National Geographic* cover story, the use of art was shown to make significant improvements in overall coping for treating traumatic brain injuries. Service members were encouraged to create a "mask" that they believed depicted either their experience in combat or their experience reintegrating into civilian life. These masks gave credence to what actually happened from these individuals' experience. Art therapist Melissa Walker encouraged this process, as it revealed "hidden feelings"; soldiers, though, were skeptical stating, "I wanted no part in it. . . . Well, I was ignorant, and I was wrong because it is great. I think this is what started me kind of opening up and talking about stuff and actually trying to get better."[36] Art reached into an abyss to name raw emotions that perhaps could not have been met merely with words. The community is encouraged to share our stories in the manner in which the narrative insists.

To conclude this section, O'Brien puts it best: "In the end, of course, a true war story is never about war. It's about the special way that dawn spreads out on a river when you know you must cross the river and march into the mountains and do things you are afraid to do. It's about love and memory. It's about sorrow. It's about sisters who never write back and people who never listen."[37] After an individual—or individuals—share their story, it is time to pray.

Prayers

Phillip wants to pray again. He was clear about that. Similar to his reflection about picking up the guitar again after his season away, there is a present numbness and new calluses needing to form before new habits can cement. The chaplain and the counterhegemonic community pray for those who cannot pray. The community is present *in* prayer. The prayer section contains an individual prayer that the participants can recite aloud or internally and also a corporate prayer to acknowledge the societal implications of these wars. Built into this time is a time for silence and solitude. Within this silence, participants can offer personal prayers, continue to recite the corporate prayer, meditate, or utilize the time in ways that are meaningful for them. First, the individual prayer:

> *Facilitator:*
> Holy One,
> We come before you, thankful for another day
> to live into your glory.
> We pray that we would realize this gift.
> In those moments, though,
> of suffering, pain, and death
> we look for you;
> we look for your Son.
> We are comforted, that the Triune God suffers too:
> Creator, Liberator, and Healer.
> Search us, sustain us.
> Examine what needs to be examined;
> lament what needs to be lamented;
> heal what needs to be healed.
> For the grace and peace that surrounds us,
> we give You thanks. *Amen.*

At the conclusion of the individual prayer, a time of silence ensues. Following silence, the corporate prayer is next. A corporate prayer is a vital element of the Liturgy of Solidarity as it provides the space for the veterans *and* the community to acknowledge the presence of war, moral injury, and the ramifications of those realities. The prayer acknowledges that it is *not* merely a veteran "problem," but it is a situation in which humanity must move forward toward reconciliation and healing together.

> *Facilitator:*
> Let us pray.
> Facilitator and Community:
> We come before you today a broken people.
> Some broken by the memories of actions we cannot undo;
> others broken by an inaction we never knew we asked you to do.
> We have gone to wars;
> we have sent you to wars.
> We all come home.
> We all suffer;
> we beseech You, come.
> Grant us the courage to stand up for what we believe;
> forgive us when we fail to do so.
> Mend our divisions;
> may we care for our neighbors.
> As the One who suffers with,
> surround us.
> In Your Grace and Peace, we pray. *Amen.*

After a time of prayer—both individual and corporate—the liturgy is prepared to begin confession and forgiveness.

Absolution: Confession and Forgiveness

The United Church of Christ's *Book of Worship* begins the Order for Corporate Reconciliation with a prayer of confession that prepares the entire community for confession:

> *Facilitator and Community:*
> Holy One,
> our sins are too heavy
> to carry, too real to hide,
> and too deep to undo.
> Forgive what our lips tremble to name,
> what our hearts can no longer bear, and what has
> become for us a consuming fire of judgment.

> Set us free from a past
> that we cannot change;
> open to us a future in which
> we can be changed;
> and grant us grace to grow
> more and more
> in your likeness and image;
> through Jesus Christ our Savior. *Amen.*[38]

Confession is a spiritual discipline and practice that enables the veteran *and* civilian an opportunity to account for the actions that took place within their MIE. It is crucial to stress that from a caregiving perspective, a chaplain should not offer confession *as a means of forgiving "sin" and therefore placing blame*. Traditional theological categories of sin—in which actions are done *to* God—are not necessarily helpful, as they can continue to isolate the veteran from the community. Process theologian Marjorie Suchocki, reminds us, "We are ourselves corporately responsible for the societies we create and the ill effects they engender."[39] Sin, too, must be reconceptualized within a communal understanding. There is a communal responsibility for these wars and, therefore, a communal responsibility for reckoning with their consequences. Sin "always affects society."[40] With that expansion in mind, throughout this liturgy I use "transgression" instead of "sin." In Jinkerson's symptomology from chapter 2, guilt and shame are primary responses to a MIE; therefore, the confession and forgiveness within the liturgy is structured around this reality and intentional in the wording to avoid further traumatizing.

All of this is not to say that there is not a place for taking accountability for the MIE and proactively taking steps to mend the pain imparted or experienced. With that said, the liturgy moves into a time of confession. In chapter 2 I touched on Harris's Building Spiritual Strengths (BSS) as a spiritual intervention model. BSS is of particular interest because within steps 6 (explore theodicy [spiritual explanations for suffering]), and 7 (explore and reframe forgiveness of self and others), the participant has the opportunity to explore forgiveness for others, self, or a Higher Power. The ongoing forgiveness is appropriate because "forgiveness is discussed as an ongoing process of maintaining an appropriate relationship with the one in need of forgiveness, rather than a 'forgive and forget' approach to resolving conflict."[41]

> *Facilitator:*
> In the Name of the triune God:
> Creator, Liberator, and Healer.
> *Facilitator and Community:*
> Amen.

Facilitator:
Let us pray:
Facilitator and Community:
Holy One,
from whom comes all holy desires and just works,
breathe into our hearts by your Holy Spirit
the gift of obedient faith,
that we, knowing your will,
may treasure these words in our minds and hearts
and may in all things love and serve you. *Amen.*
Facilitator:
Since we have a
great high priest
who has passed through the
heavens, Jesus, the only one
begotten by God,
Facilitator and Community:
Let us then
with confidence
draw near to the
throne of grace,
that we may
receive mercy
and find grace
to help in time of need.
Facilitator:
Let us confess our transgressions
Facilitator and Community:
Have mercy on us, Holy One,
according to your steadfast love;
according to your abundant mercy
blot out our transgressions.
Create in us
a clean heart,
Restore to us the joy of your salvation
through Jesus Christ. *Amen.*

The community now observes a time of silence. This silence is utilized for personal confession, prayer, meditation, or in whichever manner the participant feels is appropriate. When the veteran is ready, forgiveness is imparted, as a way of absolving guilt and shame. The liturgy, therefore, transitions into a time of assurances of pardoning from our transgressions.

Facilitator:
Jesus said to a sinner:

> Where are your accusers?
> Has no one condemned you?
> *Facilitator and Community:*
> Neither do I
> condemn you;
> go, and sin no more.[42]

Prayer of Peace (Benediction)

Following the sharing of stories, prayers, confession and absolution, the official liturgy concludes with a Prayer of Peace, modified from the *Book of Worship* of the United Church of Christ. The prayer focuses on peace as this is the message that the counterhegemonic community imparts to each member and their community. The peace is not merely the absence of war, but it is the peace that is on offer that a liberative God is co-suffering in those very moments with us. Therefore, we conclude:

> *Facilitator:*
> Go forth into the world in peace;
> be of good courage;
> hold fast to that
> which is good;
> render to no one evil for evil;
> strengthen the fainthearted;
> support the weak;
> help the afflicted;
> honor all people;
> love and serve God,
> rejoicing in the power of the Holy Spirit. *Amen.*[43]

The Liturgy of Solidarity can take place in any setting. The liturgy is malleable enough to be done quickly or with even further intentionality. The point of the liturgy, though, is to come together with reverence for the memories of one another's war experiences. The way we as a society begin to oppose dominant-hegemonic ideological "readings" of war stories is through the intentionality solidarity.

Beyond the liturgy, it may also be therapeutic for the community to proactively take steps to atone for MIEs via public—or community—service. Public service affords the veteran and the community time together supporting and serving the community in a broader cause. Veterans who miss the mission focus of military life could benefit from working together for a new common cause. Further, as Moon elaborates, "The simple task of building a house or cleaning up the neighborhood can provide inviting, nonjudgmental, and mutually empowering experiences for all participants."[44] Timing wise, these events could be scheduled around major holidays (Veterans Day, Memorial Day, or the Fourth of July), or around specifically meaningful days for the community (e.g., in remembrance of a fallen battle buddy). Even without the intentionality of timing, the service offered can be offered at any point. Beyond the liturgy, there is still an important place within the healing process for pastoral counseling, to which we now turn.

As a military chaplain, one of my overall goals is, as pastoral theologian Karen Scheib reminds us: proclaim God's solidarity with us in our suffering.[45] This theological reflection *is* possible. Returning, once again, to Dietrich Bonhoeffer and to liberation theologies offers inroads for the implications of a suffering God.

ONLY A SUFFERING GOD CAN HELP: BONHOEFFER AND LIBERATION THEOLOGIES

In chapter 5 I noted that Bonhoeffer's theology of sociality impacts his entire life's work. Also, within chapter 5's abbreviated biography, I stopped at the Nazi Gestapo's closing of Finkenwalde, the Confessing Church's seminary. Finkenwalde closed its doors at the end of 1937. Bonhoeffer would not be arrested until 1943, so he continued his work of resistance. This point in Bonhoeffer's life was trying. Although Dietrich was eligible for military service—born between 1906 and 1907—he did not want to join. Avoiding military service would prove difficult. Dietrich decided, in November 1940, to accept military service. Dietrich accepted a position in the Abwehr, as this would keep him a safe distance from a frontline military call-up or arrest. The Abwehr was, loosely, the counterintelligence agency. The Abwehr had long been involved in plans to overthrow or assassinate Hitler. Bonhoeffer portrayed himself as a pastor in the Abwehr as one unfamiliar with military duties. The Gestapo ultimately arrested Bonhoeffer in March 1943 on charges of "subversion of the armed forces," and imprisoned him in Tegel. He was later sent to Buchenwald concentration camp in February 1945, moved to Flossenbürg in April, and executed on April 9.

While in prison, Bonhoeffer wrote letters, book fragments and proposals, wedding sermons, and devotional material later compiled by lifelong confidant Eberhard Bethge and posthumously published as *Letters and Papers from Prison*. These letters were in correspondence to his family and most importantly to Bethge. Bonhoeffer's prison *Letters* offer a final opportunity to decipher how to live in community in a society that has "come of age," and how a suffering God can help reintegrate veterans experiencing moral injury. A world that has "come of age" is the culmination of a shift—one that Bonhoeffer signaled in *Sanctorum Communio*—in which humanity has complete objective autonomy. Humanity does not need the divine to solve problems, since reason now resides completely with the individual person. Bonhoeffer's world "come of age" should not necessarily be read in a pejorative manner: Bonhoeffer is not elevating one type of position over the next. He is more interested in humanity's response to existential questions. Further, humanity living in a world "come of age" takes responsibility for one's actions, and do rely on outside resources. Religion, then, represents a God that is banished to the periphery and becomes a *"deus ex machina"* swooping in to solve humanity's problems.[46] For Bonhoeffer, the societal gravitation toward secularization (and therefore the movement away from the divine) is not concerning because it emphasizes that Christianity is *of this world* and committed to the lived experience of this world.

Instead of a *deus ex machina*, Bonhoeffer's Jesus represents a God of "powerlessness and suffering; only the suffering God can help."[47] Bonhoeffer, asking a similar question to Phillip's, wants to know about those "anxious souls" that will ask "what room there is left for God now."[48] Bonhoeffer, ultimately, agrees with Phillip—the arbitrary God that swoops in to protect God's people *abandoned* Phillip. There is a twist before Bonhoeffer's proposal, however. Bonhoeffer states that God "compels" humanity to embrace—live into—God's absence. It is the same absence Jesus felt on the cross and recorded in the Gospel according to Mark: "At three o'clock Jesus cried out with a loud voice, 'Eloi, Eloi, lema sabachthani?' which means, 'My God, my God, why have you forsaken me?'"[49]

For Bonhoeffer, it is on the cross that God has allowed Godself an alienation from the world "come of age." However, that alienation and death on the cross is precisely where hope is cultivated. The hope is cultivated in two instances: first, the omnipotence of Phillip's God that disappeared during reintegration is resurrected into a God that, in weakness and suffering, is in solidarity with him. The Apostle Paul says as much:

> For the message about the cross is foolishness to those who are perishing, but to us who are being saved it is the power of God. For it is written, "I will destroy the wisdom of the wise, and the discernment of the discerning I will thwart." Where is the one who is wise? Where is the scribe? Where is the debater of this age? Has not God made foolish the wisdom of the world? For since, in the wisdom of God, the world did not know God through wisdom, God decided, through the foolishness of our proclamation, to save those who believe. For Jews demand signs and Greeks desire wisdom, but we proclaim Christ crucified, a stumbling block to Jews and foolishness to Gentiles, but to those who are the called, both Jews and Greeks, Christ the power of God and the wisdom of

God. For God's foolishness is wiser than human wisdom, and God's weakness is stronger than human strength.[50]

God, through the event of the cross and resurrection, is able to remain in solidarity with the pain, desolation, and abandonment of humanity. Bonhoeffer is not alone in this assertion. Black liberation theologians also note that Jesus, as the crucified one, is in solidarity with the oppressed. James Cone makes this point most emphatically in two of his works, 1975's *God of the Oppressed* and 2011's *The Cross and the Lynching Tree*.[51]

In *The Cross and the Lynching Tree*, Cone attempts to rehabilitate the cross and its centrality for the Black community. Cone methodically connects the cross with the lynching era, from 1880 to 1940, in which white people executed thousands of Black people. Even those white people who did not *actually do* the lynching participated in the spectacle of lynching by attending the "event": upward of 20,000 people attended. For Cone, the terror of lynching represents a similar terror of crucifixion. Cone states, "like Jesus, hanging on a cross, this nameless black victim, hanging on a Georgia tree, was left to a shameful death."[52] Of course, the excruciating pain was administered to the individual on the apparatus, but the terror was also for the audience: this is what happens to political dissidents. The apparatuses were visual reminders. To live in the reality of the lynch mobs, demanded a need for a theological solidarity with the cross. Therefore, following the same ontological argument as Cone's *God of the Oppressed*, Cone argues that God not only identifies *as* Black, God further identifies with the lynched because God, too, was lynched. More specifically, Jesus was hated, arrested, subject to a phony trial, humiliated, and executed. Jesus, therefore, has sympathy, as he is literally "suffering-with."[53]

Phillip felt "thrown into the world" once he got home, and a suffering God knows that and is with him in that. The previous understanding of God—Bonhoeffer's *deus ex machina*—understands power differently. The God that Bonhoeffer and the various liberation theologians are advocating has power in suffering. Cone speaks to this paradoxical reality of God's power noting that its paradox comes via a dialectic, and an example par excellence of this dialectic: "A symbol of death and defeat, God turned [the cross] into a sign of liberation and new life."[54] Therefore, the dialectic must remain: The suffering and the helping; God's weakness and God's strength; God's incarnation in Jesus and God's transcendence. Maintaining the dialectic is what holds the potential for transformation. Jesus's alienation on the cross matches Phillip's alienation upon reintegration.

Further, this alienation may continue; a Liturgy of Solidarity may take time to truly mend the wounds of war. Therefore, the chaplain's role through liturgy, spiritual care and counseling, or solidarity through accompaniment is to continue to show up; to continue to represent a God of strength through God's weakness. In other words, change and healing are slow processes, and through the long unraveling of time the chaplain and the community's role are to sustain those individuals on that journey. There is grace in this each step of this journey; from the first telling of a war story to the umpteenth "thank you for your service," liberation is ongoing.

Second, hope is cultivated through a suffering God in that humanity is now empowered to take accountability and responsibility for caring for our neighbors. For the development of a counterhegemonic community, a suffering God has implications for the community as well. God is able, through Jesus, to be with humanity, truly in their lived experiences. Bonhoeffer elaborates this as, "His 'being there for others,' maintained till death, that is the ground of his omnipotence, omniscience, and omnipresence."[55] The church, the community, or the small counterhegemonic communities I am advocating participate in this presence. The calling, then, is one of solidarity with one another. In Latinx liberation theologies, this solidarity is experienced as a strengthening in our suffering not only for others but also for ourselves. This solidarity is immensely hopeful. God went through it, Jesus went through it, and so we can go through it. Humanity experiences God's transcendence through our neighbors, those individuals who are co-suffering and co-journeying in solidarity. God is in *lo cotidiano* (the everyday).[56] Jesus is rooted to the veterans experiencing moral injury. The "church is the church only when it exists for others," as Bonhoeffer reminds us.[57] This is elaborated as taking on the problems of the world, through service to one another.

To conclude, Bonhoeffer's suffering God relates to the world "come of age." Writing to Bethge, he notes:

> During the last year or so I've come to know and understand more and more the profound this-worldliness of Christianity. The Christian is not a *homo religiosus*, but simply a man, as Jesus was a man. . . . I discovered later, and I'm still discovering right up to this moment that is it only by living completely in this world that one learns to have faith. . . . By this-worldliness I mean living unreservedly in life's duties, problems, successes and failures, experiences and perplexities. In so doing we throw ourselves completely into the arms of God, taking seriously, not our own suffering, but those of God in the world—watching with Christ in Gethsemane.[58]

This project has been an attempt at taking seriously the suffering of "those of God in the world." The narratives in this book are of suffering and alienation, but a suffering God can help. Before providing closing remarks on this study, I want to offer best practices of theological reflection and interpretation on when I would have introduced a suffering God.

BEST PRACTICES AND THEOLOGICAL INTERPRETATION

I began to officially introduce verbiage of a suffering God to Phillip. However, it fell outside our time together, but I want to briefly address how introducing a new theological interpretation is possible in a caregiving relationship, and what considerations I take into account when deciding whether or not to address it. Within Osmer's normative task of practical theology, a practical theologian brings "ethical norms," values, and presuppositions into each situation—or, for me, into each pastoral counseling session.[59] Within my own theological reflection, I bring the values of a suffering God into the room with me. Those values are supported by the various liberative praxis methods that strengthen my theological framework: God's solidarity with the marginalized, God's presence in history in peoples' pain, and the radiant possibility of hope moving forward toward restoration. Introducing these new ideas to Phillip originated out of discerning my own curiosity and how I understood Phillip storying his narrative. In this vignette, I started to address my curiosity, slowly, by asking:

Chaplain: Do you think God was affected by the accident [referencing the MIE]? Does your pain matter to God?

Phillip: What?

Chaplain: Was God sad, moved, angry, upset, or some other emotion after that day on the road?

Phillip: I hope he is sad.

Chaplain: Tell me about what it might mean for God to be sad about our losses.

Phillip: He would have to be there to *feel* it, you know?

Chaplain: Yes, I do. I agree.

Phillip: I could get around to that. . .

I was introducing a reflection to gauge its significance for Phillip; essentially, would a suffering God resonate? After I asked Phillip about God's emotions, if he did not engage the idea or if he provided a different understanding of how God was affected by the MIE, I would not have pursued the suffering God motif further. Continuing to import my ethical and theological norms into a spiritual care and counseling session, while presupposing that Phillip is not bringing his own norms, is unethical and anathema to the methods of a liberative praxis of pastoral care.

FUTURE RESEARCH OPPORTUNITIES

It is possible to oppose dominant ideologies. Individuals cannot do it alone, but rather, a community is needed in which each member is committed to opposing dominance. This opposition will take all the intellectual and creative energy of a community. This opposition is a slow process as well. It takes time to build trust and rapport in communities, and the task asks for a vulnerability to share one's story—perhaps a story that has not been shared with anyone previously.

With that in mind, there are two areas in which this study can elicit further reflection and study. First, theologically, this project was limited to the reintegration experience of various strands of Christianity, whether nondenominational evangelical Christian or even a moralistic therapeutic deism permeated by quasi-Christian values. Much of the framework and the Liturgy of Solidarity, then, is built around a Christian ecclesiology of religious communities. Even within Christianity there is enough variance to necessitate different practices and commitments of a counterhegemonic group. Broaden these communities beyond Christianity, and the differences are more pronounced. However, I still contend that practices of reconciliation via prayer, sharing mutual concern, sharing one's story, and the place of confession is still available outside Christianity as well. With that said, the tension between spontaneity and order that I addressed above is all the more important at this point; namely, the liturgy can—and should—be adapted to fit the religious and spiritual needs of those within the counterhegemonic community.

Second—perhaps naïvely—what becomes the long-term support for the wars in Iraq and Afghanistan? Our collective discourse needs differentiation between an anti-war position that obfuscates the lived reality of veterans and an anti-war position that is "oppositionally" counterhegemonic. Would this platform defeat itself without the support of veterans who are not anti-war? How can LaMothe's "unconventional warriors" be expanded as a platform for reintegration?

Finally, it is difficult to imagine ongoing support of these wars, if one is able to hold in tension the traumatic tolls they entail. PTSD, moral injury, increasing suicide rates, and a globe reeling from America's hegemonic foreign policy are all alarming. The counterhegemonic communities I am proposing have the responsibility to broaden the intellectual and creative imagery for ending these wars. After ideological imagery shifts, awareness would be raised concerning the effects of these wars. Military chaplains *as clergy* are on the front lines of this shift, this *war of position*. In the caregiving relationship, though, on a smaller scale, military chaplains assist in bringing veterans—completely—home. Speaking to this, Chaplain Herman "Herm"

Keizer Jr., original codirector of the Soul Repair Center with Rita Brock, describes the role of a military chaplain:

> I am constantly amazed that the people to whom chaplains minister continue to trust us with their stories, opening themselves in conversations both painful and prideful; complex in their simplicity and simple in their complexity; wonderful and terrible stories that reveal the vulnerability and resiliency of the human spirit. They share their life stories with us and welcome us into their narrative, allowing us to interpret their stories with them.[60]

Chaplains live in the ambiguity. Yes, this is a project about critiquing and challenging ideological imagery that complicates reintegration. However, on the personal, caregiving level, military chaplains must not forget that it is a hallowed ground in which veterans "share their life stories with us and welcome us into their narrative, allowing us to interpret their stories with them," which continues to cause pause for reflection; interpreters of stories are needed.

These small, yet life-giving, counseling sessions allow veterans to re-author a debilitating narrative that has caused personal, familial, and societal struggles. From these "living human documents" care is reconceptualized on a broader scale, impacting families, religious communities, and society at large. Change comes slowly, but change is possible.

NOTES

1. Some of the theoretical analysis in this chapter was developed in, "Veteran Support and Antonio Gramsci: Counterhegemony as a Pastoral Theological Intervention," *Journal of Pastoral Theology* 30, no. 3 (2020): 207–221.
2. Lartey, *In Living Color*, 57–58.
3. Adam David Morton, *Unraveling Gramsci: Hegemony and Passive Revolution in the Global Political Economy* (New York: Pluto Press, 2007), 35. Further, Hall discusses the reticence to use Gramsci as an "Old Testament prophet who, at the correct moment, will offer us the consoling and appropriate quotation." Hall's point is that we must not conflate our epoch—or "conjuncture" in Hallsian terms—with Gramsci's but rather focus on asking better questions of our current situation. Stuart Hall, "Gramsci and Us," *Marxism Today* (June 1987): 16.
4. Moon, *Warriors Between Worlds*, 84.
5. Hall, "Encoding/Decoding," 138.
6. Antonio Gramsci, *Prison Notebooks*, ed. and trans. Joseph A. Buttigieg, vol. III (New York: Columbia University Press, 2007), 168. This is also sometimes referred to as the "military metaphor."
7. For Trotsky, in particular, the metaphor is *not* a metaphor. He states, "Insurrection, armed insurrection . . . was inevitable from our point of view. It was and remains a

historical necessity in the process of the people's struggle against the military and police state." Leon Trotsky, *1905* (New York: Vintage Books, 1971), 394, quoted in Daniel Egan, "Rethinking War of Maneuver/War of Position: Gramsci and the Military Metaphor," *Critical Sociology* 40, no. 4 (2014): 528.
8. Roland Boer, *Criticism of Heaven: On Marxism and Theology*, Historical Materialism Book Series (Chicago: Haymarket Books, 2009), 270.
9. Gramsci, *Prison Notebooks*, III, 169.
10. Gramsci, *Selections*, 334.
11. Stuart Hall, "Domination and Hegemony," in *Cultural Studies 1983: A Theoretical History*, ed. Jennifer Daryl Slack and Lawrence Grossberg (Durham, NC: Duke University Press, 2016), 178.
12. Boer, *Criticism of Heaven*, 238.
13. Antonio Gramsci, *Prison Notebooks*, ed. and trans. Joseph A. Buttigieg, vol. I (New York: Columbia University Press, 1992), 163.
14. Gramsci, *Prison Notebooks*, I, 454. Buttigieg also notes that Gramsci had André Philip's 1927 text *Le Problème ouvrier aux États-Unis* with him in prison. This text detailed the role of the clergy in the social movements mentioned above.
15. Kate Crehan, *Gramsci's Common Sense: Inequality and Its Narratives* (Durham, NC: Duke University Press, 2016), 33.
16. Stuart Hall, "The Formation of Cultural Studies," 11–12.
17. Gramsci, *Prison Notebooks*, I, 234.
18. Antonio Gramsci, *Further Selections from the Prison Notebooks*, ed. and trans. Derek Boothman (Minneapolis: University of Minnesota Press, 1995) 98.
19. Boer, *Criticism of Heaven*, 247.
20. Antonio Gramsci, *Prison Notebooks*, ed. and trans. Joseph A. Buttigieg, vol. II (New York: Columbia University Press, 1996), 222.
21. Ibid., 224.
22. Ibid., 223.
23. Boer, *Criticism of Heaven*, 258.
24. Stephen Pattison, *Pastoral Care and Liberation Theology*, 2nd ed. (London: SPCK, 1997), 48.
25. Watkins Ali, *Survival and Liberation*, 120.
26. Waggoner, "Taking Religion Seriously," 718.
27. Ibid., 717.
28. Ibid., 717–718.
29. David W. Peters, *Post-Traumatic God: How the Church Cares for People Who Have Been to Hell and Back* (New York: Morehouse Publishing, 2016), 105.
30. Kinghorn, "Combat Trauma," 71.
31. Litz et al., "Moral Injury," 701–703.
32. Elaine Ramshaw, *Ritual and Pastoral Care* (Philadelphia: Fortress Press, 1987), 24–25.
33. Ibid., 25.
34. Tim O'Brien, *The Things They Carried* (New York: Mariner Books, 2009), 78.
35. Ibid., 65, 74.

36. Caroline Alexander, "The Invisible War on the Brain," *National Geographic*, February 2015, 44.
37. O'Brien, *The Things They Carried*, 81.
38. Adapted from United Church of Christ, *Book of Worship* (Cleveland: United Church of Christ Press, 2012), 271. In the spirit of UCC polity, I have revised to meet my context.
39. Marjorie Hewitt Suchocki, *God, Christ, Church: A Practical Guide to Process Theology* (New York: Crossroad Publishing Company, 1995), 15.
40. Ada María Isasi-Díaz, *Mujerista Theology: A Theology for the Twenty-First Century* (New York: Orbis Books, 1996), 90.
41. Harris et al., "Trauma-focused Spiritually Integrated Intervention," 428.
42. Adapted from United Church of Christ, *Book of Worship*, 277–285. In the spirit of UCC polity, I have revised to meet my context.
43. Ibid., 288. In the spirit of UCC polity, I have revised to meet my context.
44. Moon, *Warriors Between Worlds*, 101–102.
45. Karen D. Scheib, *Pastoral Care: Telling the Stories of Our Lives* (Nashville: Abingdon Press, 2016), 29.
46. Bonhoeffer, *Letters and Papers*, 361. Emphasis in original.
47. Ibid.
48. Ibid., 360.
49. Mark 15:34.
50. 1 Cor 1:18–25.
51. James H. Cone, *The God of the Oppressed*, rev. ed. (New York: Orbis Books, 1997); James H. Cone, *The Cross and the Lynching Tree* (New York: Orbis Books, 2011).
52. Cone, *Cross and the Lynching Tree*, 77.
53. Paul S. Fiddes, *The Creative Suffering of God* (Oxford: Oxford University Press, 1988), 16.
54. Cone, *Cross and the Lynching Tree*, 156.
55. Bonhoeffer, *Letters and Papers*, 381.
56. Miguel A. De La Torre, *The Politics of Jesús: A Hispanic Political Theology* (New York: Rowan & Littlefield, 2015), 86.
57. Bonhoeffer, *Letters and Papers*, 382.
58. Ibid., 369–370.
59. Osmer, *Practical Theology*, 131.
60. Brock and Lettini, *Soul Repair*, 79–80.

Appendix

NOTE ON RESEARCH DESIGN

This project utilizes an "collective case study," as I brought my own research questions to the multiple case study.[1] Case study understands research questions as "issues." Utilizing the Greek alphabet, researchers designate these with the Greek I (iota). The issues are what compel the researcher to research a particular issue. Issues assist the researcher in expanding the living human web woven within each case study. Here are five preliminary issues I brought to this project:

- I_1: How is the combat reintegration experience unique for reserve component veterans?
- I_2: What is the relationship between being a reservist and reintegrating with an MIE?
- I_3: How does the spatial and temporal proximity of veterans to civilians impact reintegrating with an MIE?
- I_4: What is the relationship between a veteran and family and friends post-deployment?
- I_5: How do dominant ideologies impact reintegration with an MIE?

There are some noteworthy misconceptions about case study. One misconception of the methodology concerns the generalizability gained by case study. However, case study provides "a powerful and complex form of knowing, rooted in context and interpretation. The potential troubles with cases cannot be eliminated, yet people of faith and a suffering world urgently need the practical wisdom afforded by cases."[2] First, it is important to note, as Linda Dale Bloomberg and Marie Volpe have, that for case study, "generalizability is not the goal, but rather *transferability*—that is, how (if at all) and in what ways understanding and knowledge can be applied in similar contexts and settings."[3] A case study becomes transferable through the practical theological task of "thick description." This thick description "will provide the basis for a qualitative account's claim to relevance in some broader context."[4] Bent Flyvbjerg helpfully notes that a case study can enable a researcher to reach conclusions in that "if this is (not) valid for this case, then it applies to all (no) cases."[5] This is an important critique of case study, and I agree with Flyvbjerg that transferability is possible. Within case study methodology more broadly, Robert Stake's work shows how what he refers to as a "generalization"—or transferability—is possible. Stake offers a "naturalistic generalizability" that comes from human experience: readers are able to make connections between one case and various life experiences either directly or indirectly.[6]

Therefore, as with Sigmund Freud's oeuvre, for example, theories are developed out of a well-chosen and articulated case study. The process of developing a thick description is committed to the appropriate and ethical use of the case study data. Case study refers to this process as "triangulation." Triangulation is committed to ensuring, with as much accuracy as possible, an ethical study of a phenomenon. Case study triangulation attempts to provide a "substantial body of uncontestable description."[7] The outcome of this is a detailing of events that any observer would have noticed. This is fairly straightforward, and it is followed by interpretation. Ambiguous or contested interpretations can be supported with more "uncontestable description," but also through other strategies. With respect to validation, this project utilized, in addition to triangulation, four of John Creswell's strategies: "peer review or debriefing," "clarifying researcher bias," "member checking," and "rich, thick description."[8]

For my peer review and debriefing stage of data collection and analysis, I debriefed my emotions and methods with selected "devil's advocates."[9] My devil's advocates consisted of four individuals, all PhD colleagues, who served as my peer review team: three women (one Latinx Christian feminist practical theologian, one Korean American Christian pastor and practical theologian, one Caucasian feminist practical theologian of multiple religious belonging) and one man (European American Buddhist practical theologian). We met twice a week throughout the data collection and project writing phase. I presented research, emotions, and early theoretical analyses. Much of my analyses were critiqued and sharpened through their collaboration.

As I cannot escape my subjectivity, I shared my researcher bias with my participants and in chapter 1 and other parts of this project. I have noted my own situatedness and living human web connectivity that impacts how

I understand veteran reintegration. Finally, I did member checking by consulting with my participant about whether I was presenting their experience accurately. I constructed my research design in such a way that the beginning of interviews two and three began with a check-in, as this analysis determined how subsequent questions were asked. I did find that I made numerous assumptions about pieces of data and how they correlated to reintegration. When critiqued, I asked clarifying questions to correct and strengthen my understanding of the thick description. My participants were offered the chance to review the transcripts of interviews via these options: (1) read the transcript themselves, (2) have it read to them by the principal investigator, or (3) listen to the recorded interview with me present. Once again, this review was an opportunity for the participant to make corrections and clarify intended communication. With all of these theoretical and validity underpinnings laid out, I will next describe my research design.

My early phone calls and searches centered on local moral injury support groups, the Moral Injury Association of America, and the Kansas City VA Hospital. The Moral Injury Association of America quickly responded and, while noting the importance of the research, informed me that due to the sensitivity of the program I could not participate in nor observe any of their support groups; therefore, I could not invite any of their support group participants to participate in this research project. To some degree, I anticipated this level of privacy and implementation of Health Insurance Portability and Accountability Act (HIPAA) laws.

However, I was somewhat surprised by the lack of support to further the research on moral injury. In some respects, the lack of support matches my frustrations listed in chapters 1 and 2: as moral injury is still, for all intents and purposes, in its infancy, it continues to exist as a phenomenon of differential diagnosis. Conversations center on the work of Litz or Shay, and as the wars in Iraq and Afghanistan rage on toward two decades, seldom are broader cultural questions asked.

As with the Moral Injury Association of America, my request to recruit participants directly from the Kansas City VA was declined. I met with the Kansas City VA's chief of chaplains numerous times via phone calls and in person about this project and what I sought in terms of participants. The chief of chaplains helpfully explained that the VA is a difficult institution to get to facilitate outside research (for similar reasons to what I encountered with the Moral Injury Association of America). What this arduous process provided, though, were more community gatekeepers to contact and, more important, an emerging path to follow.

During those early months of recruitment and the lessons learned, I decided to move away from a reliance on institutions, such as the VA, for another critical reason: many of the VA's support groups for moral injury are filled entirely by active-duty veterans. As this project is also interested in the Reserve component, and what I found through my recruitment process further undergirds the necessity for this project. Reserve component veterans continue to experience their military service as something *other*, something less than, and this othering comes about through ideological interactions with civilians. In response to the first question of the first interview, Phillip immediately detailed to me why his service did not matter, and he stated that he is "just like any other Reserve soldier." This was not a false modesty; rather, his dismissive tone communicated everything and confirmed much of my own experience as a reservist over twelve-plus years. In the military, there is a dichotomy: active duty and then everyone else. How do Phillip and Angela, then, reintegrate within a society in which the VA—*the* veteran healthcare institution—does not implicitly treat their service on a similar level as their active-duty counterparts? Returning to recruitment, the movement away from institutions helped me locate therapists and military chaplains who are providing moral injury support groups or pastoral counseling with veterans experiencing the effects of MIEs in a both an Active Duty and also a Reserve component context.

Interview Structure

The interviews were "the main road to multiple realities."[10] To maintain self-reflexivity in my interviewing I built a period of reflexive analysis into my data collection. I formally interviewed my participants three times. The first interview was a general semi-structured interview lasting sixty minutes. During that interview I asked questions centering on the essence of reintegration. The case study ideas I described above cultivated these questions, which, to recap, were as follows:

I_1: How is the combat reintegration experience unique for reserve component veterans?
I_2: What is the relationship between being a reservist and reintegrating with an MIE?
I_3: How does the spatial and temporal proximity of veterans to civilians impact reintegrating with an MIE?
I_4: What is the relationship between a veteran and family and friends post-deployment?
I_5: How do dominant ideologies impact reintegration with an MIE?

These experiential questions were broken into subset questions on how the participants' (1) military unit reintegrated them, (2) how family experienced them, and (3) how they felt within broader society. I then sorted and analyzed these interviews.

Weeks one, three, and five were formal interview weeks. None of these interviews exceeded sixty minutes. Built into my research design was time to sit with the data. Linda Finlay notes the importance of "dwelling" with data to discover the multivalent meaning within the raw material.[11] The dwelling stage offered time for a mutual indwelling during which I reflected through journaling on my own experiences of deploying and reintegrating. The mutual indwelling put me *experientially* in combat with my participants. When Phillip or Angela described mountain ranges in southern Afghanistan, I recalled the mountain ranges throughout Kandahar, where I had been stationed. I listened to the music my participants listened to. I spent time with the photos my participants shared. I read the journal entries my participants shared. I wanted to immerse myself in my participants' phenomenological "lifeworlds." The interviews came to life, so to speak, as I dwelled with them. I dwelled with each interview before transcribing it, after the transcription, before coding, and after the coding. Each of these steps was an opportunity for empathy with the story I was learning.

Research Journal

I maintained a research journal for each interview and interaction. Almost immediately following each interview, and not exceeding twenty-four hours post-interview, I journaled my reflections and interpretations into a research journal to note researcher bias and overall impressions of my time. I began coding after journaling, and I spent a week sitting with the transcripts and my journal. In other words, weeks two, four, and six were reserved for coding and further analysis. The journal proved to be a site of exploration where I could process frustrations with the mechanics of the interview process and frustrations with the subject matter overall. What this allowed for was a space to process and check my subjectivity in a separate location from direct participant interaction. Some early theological and theoretical insights came from these rough journaling periods. Finally, I analyzed my journals for themes that were coming up for me, and I cross-referenced them with the coded themes in the actual interview transcripts.

Additional Data Collection

It was advantageous to analyze other sources of data besides spoken-word interviews. For example, as chapter 4 highlighted, music was a vital coping mechanism for Phillip. Therefore, during interview two, I learnt artists, specific songs (and lyrics), and themes to study. This, in conjunction with the transcribed interview, developed a more complete picture of his deployment experience. When my participants shared stories of Iraq or Afghanistan, I would ask for photos. Lisa graciously shared multiple journal entries of key points from her deployment as well as her reintegration entries stateside. Once I received pictures or journal entries, I would analyze these in my data-collecting software, NVivo.

Coding

With my participants, I tried to let the interview happen organically and not "pre-think" my coding, with the exception of the preliminary ideas listed above. I uploaded each transcript and research journal into NVivo for coding. I finalized and edited the codes down to six designated codes, or "nodes" in NVivo, and these had subcodes as well. Early coding was done through an in vivo approach, in which I took codes "from the actual language" of the interview and began shaping my interpretation of them.[12] For example, what became a dominant theme—"belonging"—was originally coded as "service didn't matter," as this was a direct quote from interviews one and two.

NOTES

1. John W. Creswell, *Qualitative Inquiry & Research Design: Choosing Among Five Approaches*, 3rd ed. (Los Angeles: Sage Publications, 2013), 99.
2. Eileen R. Campbell-Reed, "The Power and Danger of a Single Case Study in Practical Theological Research," in *Conundrums in Practical Theology*, ed. Joyce Ann Mercer and Bonnie J. Miller-McLemore (Boston: Brill, 2016), 54.
3. Linda Dale Bloomberg and Marie Volpe, *Completing Your Qualitative Dissertation: A Road Map from Beginning to End*, 3rd ed. (Los Angeles: Sage Publications, 2016), 47.
4. Ibid.
5. Bent Flyvbjerg, "Five Misunderstandings About Case-Study Research," *Qualitative Inquiry* 12, no. 2 (April 2006): 230.
6. Stake, *Art of Case Study Research*, 85–88.
7. Ibid., 110.
8. Creswell, *Qualitative Inquiry & Research Design*, 250–252.
9. Ibid., 251.
10. Stake, *Art of Case Study Research*, 64.
11. Linda Finlay, *Phenomenology for Therapists: Researching the Lived World* (West Sussex, England: Wiley Blackwell, 2011), 229.
12. Johnny Saldaña, *The Coding Manual for Qualitative Researchers*, 3rd ed. (Los Angeles: Sage Publications, 2016), 77.

Bibliography

Alexander, Caroline. "The Invisible War on the Brain." *National Geographic*, February 2015.
Alexander, D. William. "Gregory Is My Friend." In *War and Moral Injury: A Reader*, edited by Robert Emmet Meagher and Douglas A. Pryer, 197–207. Eugene, OR: CascadeBooks, 2018.
Althusser, Louis. *Lenin and Philosophy, and Other Essays*. Translated by Ben Brewster. New York: Monthly Review Press, 1971.
Asquith, Glenn H. Jr. "The Case Study Method of Anton T. Boisen." *Journal of Pastoral Care* 34, no. 2 (June 1980): 84–94.
Beaumont-Thomas, Ben. "Clint Eastwood: *American Sniper* and I Are Anti-War." *The Guardian*, March 17, 2015. Accessed April 17, 2017.
Bernstein, Jay M. "Suffering Injustice: Misrecognition as Moral Injury in Critical Theory." *International Journal of Philosophical Studies* 13, no. 3 (2005): 303–324.
———. *Torture and Dignity: An Essay on Moral Injury*. Chicago: The University of Chicago Press, 2015.
Bethge, Eberhard. *Dietrich Bonhoeffer: Theologian, Christian, Contemporary*. Translated by Eric Mosbacher, Peter and Betty Ross, and Frank Clarke. London: William Collins Sons & Co., 1970.
Bloomberg, Linda Dale, and Marie Volpe. *Completing Your Qualitative Dissertation: A Road Map from Beginning to End*. 3rd ed. Los Angeles: Sage Publications, 2016.
Boer, Roland. *Criticism of Heaven: On Marxism and Theology*. Historical Materialism Book Series. Chicago: Haymarket Books, 2009.
Boisen, Anton T. *The Exploration of the Inner World: A Study of Mental Disorder and Religious Experience*. Philadelphia: University of Pennsylvania Press, 1936.
Bonhoeffer, Dietrich. *Ethics*. Dietrich Bonhoeffer Works, vol. 6, edited by Clifford J. Green. Translated by Reinhard Kraus, Charles C. West, and Douglas W. Stott. Minneapolis: Fortress Press, 2009.
———. *Letters and Papers from Prison*, edited by Eberhard Bethge. New York: Simon & Schuster, 1997.

———. *Sanctorum Communio: A Theological Study of the Sociology of the Church*. Dietrich Bonhoeffer Works, vol. 1, edited by Clifford J. Green. Translated by Joachim Von Soosten, Reinhard Kraus, and Nancy Lukens. Minneapolis: Fortress Press, 2009.

———. *Sanctorum Communio: Eine dogmatische Untersuchung zur Soziologie der Kirche*. München: Christian Kaiser Verlag, 1960.

Brock, Rita Nakashima, and Gabriella Lettini. *Soul Repair: Recovering from Moral Injury After War*. Boston: Beacon Press, 2012.

Brown, Wendy. *Undoing the Demos: Neoliberalism's Stealth Revolution*. New York: Zone Books, 2017.

Buber, Martin. *I and Thou*. Translated by Ronald Gregor Smith. New York: Scribner Classics, 2000.

Campbell-Reed, Eileen R. "The Power and Danger of a Single Case Study in Practical Theological Research." In *Conundrums in Practical Theology*, edited by Joyce Ann Mercer and Bonnie J. Miller-McLemore, 33–59. Boston: Brill Publishers, 2016.

Carey, Lindsay B., Timothy J. Hodgson, Lillian Krikheil, Rachel Y. Soh, Annie-Rose Armour, Taranjeet K. Singh, and Cassandra G. Impiombato. "Moral Injury, Spiritual Care and the Role of Chaplains: An Exploratory Scoping Review of Literature and Resources." *Journal of Religion and Health* 55, no. 4 (2016): 1218–1245.

Chopp, Rebecca S. "Practical Theology and Liberation." In *Formation and Reflection: The Promise of Practical Theology*, edited by Lewis S. Mudge and James N. Poling, 120-138. Philadelphia: Fortress Press, 1987.

Cone, James H. *A Black Theology of Liberation*. New York: Orbis Books, 2010.

———. *The Cross and the Lynching Tree*. New York: Orbis Books, 2011.

———. *God of the Oppressed*. Rev. ed. New York: Orbis Books, 1997.

Crehan, Kate. *Gramsci's Common Sense: Inequality and Its Narratives*. Durham, NC: Duke University Press, 2016.

Creswell, John W. *Qualitative Inquiry & Research Design: Choosing Among Five Approaches*. 3rd ed. Los Angeles: Sage Publications, 2013.

De La Torre, Miguel A. *The Politics of Jesús: A Hispanic Political Theology*. New York: Rowman & Littlefield, 2015.

Denton-Borhaug, Kelly. "Like Acid Seeping into Your Soul: Religio-Cultural Violence in Moral Injury." In *Exploring Moral Injury in Sacred Texts*, edited by Joseph McDonald, 111–133. London: Jessica Kingsley Publishers, 2017.

Dingemans, Gijsbert D. J. "Practical Theology in the Academy: A Contemporary Overview." *Journal of Religion* 76 (1996): 82–96.

Doehring, Carrie. "Military Moral Injury: An Evidence-Based and Intercultural Approach to Spiritual Care." *Pastoral Psychology* 68, no. 1 (February 2019): 15–30.

Drescher, Kent D., David W. Foy, Caroline Kelly, Anna Leshner, Kerrie Schutz, and Brett T. Litz. "An Exploration of the Viability of the Construct of Moral Injury in War Veterans." *Traumatology* 17, no. 1 (2011): 8–13.

Eagleton, Terry. *Ideology: An Introduction*. London: Verso, 2007.

Egan, Daniel. "Rethinking War of Maneuver/War of Position: Gramsci and the Military Metaphor." *Critical Sociology* 40, no. 4 (2014): 521–538.

Fiddes, Paul S. *The Creative Suffering of God*. Oxford: Oxford University Press, 1988.
Finlay, Linda. *Phenomenology for Therapists: Researching the Lived World*. West Sussex, England: Wiley Blackwell, 2011.
Flyvbjerg, Bent. "Five Misunderstandings About Case-Study Research." *Qualitative Inquiry* 12, no. 2 (April 2006): 219–245.
Foucault, Michel. "The Film and Popular Memory." In *Foucault Live: Collected Interviews, 1961–1984*, edited by Slyvère Lotringer. Translated by Lysa Hochroth and John Johnston, 122–132. New York: Semiotext(e), 1996.
Frame, Tom. "Moral Injury and the Influence of Christian Religious Conviction." In *War and Moral Injury: A Reader*, edited by Robert Emmet Meagher and Douglas A. Pryer, 187–196. Eugene, OR: Cascade Books, 2018.
Geertz, Clifford. *The Interpretation of Cultures: Selected Essays*. New York: Basic Books, 1973.
Gerkin, Charles V. *The Living Human Document: Re-Visioning Pastoral Counseling in a Hermeneutical Mode*. Nashville: Abingdon Press, 1984.
Graham, Elaine, Heather Walton, and Frances Ward. *Theological Reflection: Methods*. London: SCM Press, 2005.
Graham, Larry Kent. *Moral Injury: Restoring Wounded Souls*. Nashville: Abingdon Press, 2017.
Gramsci, Antonio. *The Antonio Gramsci Reader: Selected Political Writings 1916–1935*, edited by David Forgacs. New York: New York University Press, 2000.
———. *Further Selections from the Prison Notebooks*, edited and translated by Derek Boothman. Minneapolis: University of Minnesota Press, 1995.
———. *Prison Notebooks*, edited by Joseph A. Buttigieg. Translated by Antonio Callari. 3 vols. New York: Columbia University Press, 2007.
———. *Selections from the Prison Notebooks*, edited and translated by Quintin Hoare and Geoffrey Nowell Smith. New York: International Publishers, 2014.
Green, Clifford J. *Bonhoeffer: A Theology of Sociality*. Rev. ed. Grand Rapids: William B. Eerdmans Publishing Company, 1999.
Greider, Kathleen, Gloria A. Johnson, and Kristen J. Leslie. "Three Decades of Women Writing for Our Lives." In *Feminist and Womanist Pastoral Theology*, edited by Bonnie J. Miller-McLemore and Brita L. Gill-Austern, 21–50. Nashville: Abingdon Press, 1999.
Grelle, Bruce. *Antonio Gramsci and the Question of Religion: Ideology, Ethics, and Hegemony*. New York: Routledge, 2017.
Hall, Stuart. "Culture, the Media, and the 'Ideological Effect'." In *Mass Communication and Society*, edited by James Curran, Michael Gurevitch, and Janet Woollacott, 315–348. London: Sage Publications, 1977.
———. "Domination and Hegemony." In *Cultural Studies 1983: A Theoretical History*, edited by Jennifer Daryl Slack and Lawrence Grossberg, 155–179. Durham, NC: Duke University Press, 2016.
———. "Encoding/Decoding." In *Culture, Media, Language: Working Papers in Cultural Studies, 1972–1979*, edited by Stuart Hall, Dorothy Hobson, Andrew Lowe, and Paul Willis, 128–138. New York: Routledge, 1980.

———. *Familiar Stranger: A Life Between Two Islands*. Edited by Bill Schwarz. Durham, NC: Duke University Press, 2017.

———. "The Formation of Cultural Studies." In *Cultural Studies 1983: A Theoretical History*, edited by Jennifer Daryl Slack and Lawrence Grossberg, 5–24. Durham, NC: Duke University Press, 2016.

———. "Gramsci and Us." *Marxism Today* (June 1987): 16–20.

———. "Gramsci's Relevance for the Study of Race and Ethnicity." In *Stuart Hall: Critical Dialogues in Cultural Studies*, edited by David Morley and Kuan-Hsing Chen, 411–440. New York: Routledge, 1996.

———. "Ideology and Ideological Struggle." In *Cultural Studies 1983: A Theoretical History*, edited by Jennifer Daryl Slack and Lawrence Grossberg, 127–154. Durham, NC: Duke University Press, 2016.

———. "Introducing NLR." *New Left Review* 1, no. 1 (January-February 1960): 1–3.

———. "The Neoliberal Revolution." In *Selected Political Writings: The Great Moving Right Show and Other Essays*, edited by Sally Davison, David Featherstone, Michael Rustin, and Bill Schwarz, 317–335. Durham, NC: Duke University Press, 2017.

———. "The Rediscovery of 'Ideology': Return of the Repressed in Media Studies." In *Culture, Society, and the Media: A Reader*, edited by Michael Gurevitch, 56–90. New York: Methuen, 1982.

———. "Signification, Representation, Ideology: Althusser and the Post Structuralist Debates." *Critical Studies in Mass Communications* 2, no. 2 (1985): 91–114.

Hall, Stuart, and Paddy Whannel. *The Popular Arts*. Durham, NC: Duke University Press, 2018.

Harris, J. Irene, Christopher R. Erbes, Brian E. Engdahl, Paul Thuras, Nichole Murray-Swank, Dixie Grace, Henry Ogden, et al. "The Effectiveness of a Trauma Focused Spiritually Integrated Intervention for Veterans Exposed to Trauma." *Journal of Clinical Psychology* 67, no. 4 (April 2011): 425–438.

Harris, J. Irene, Crystal L. Park, Joseph M. Currier, Timothy J. Usset, and Cory D. Voecks. "Moral Injury and Psycho-Spiritual Development: Considering the Developmental Context." *Spirituality in Clinical Practice* 2 (January 1, 2015): 256–266.

Hiltner, Seward. *Preface to Pastoral Theology: The Ministry and Theory of Shepherding*. Nashville: Abingdon Press, 1958.

Holton, M. Jan. *Longing for Home: Forced Displacement and Postures of Hospitality*. New Haven, CT: Yale University Press, 2016.

Isasi-Díaz, Ada María. *Mujerista Theology: A Theology for the Twenty-First Century*. New York: Orbis Books, 1996.

Jinkerson, Jeremy D. "Defining and Assessing Moral Injury: A Syndrome Perspective." *Traumatology* 22, no. 2 (2016): 122–130.

Johnson, Cedric C. *Race, Religion, and Resilience in the Neoliberal Age*. Black Religion, Womanist Thought, Social Justice. New York: Palgrave Macmillan, 2016.

Kassabian, Joseph. *The Hooligans of Kandahar: Not All War Stories Are Heroic*. TCK Publishing, 2017. Kindle.

Kinghorn, Warren. "Combat Trauma and Moral Fragmentation: A Theological Account of Moral Injury." *Journal of the Society of Christian Ethics* 32, no. 2 (2012): 57–74.

Klay, Phil. "The Warrior at the Mall." *New York Times*, April 14, 2018. Accessed April 25, 2018.

Krips, Henry. "Ideology and its Pleasures: Althusser, Žižek & Pfaller." *Continental Thought & Theory: A Journal of Intellectual Freedom* 2, no. 1 (June 2018): 333–367.

Kyle, Chris, and Jim DeFelice. *American Sniper: The Autobiography of the Most Lethal Sniper in U.S. History.* New York: HarperCollins, 2012.

LaMothe, Ryan. "Empire, Systemic Violence, and the Refusal to Mourn: A Pastoral Political Perspective." *Journal of Pastoral Theology* 23, no. 2 (2013): 1–26.

———. "Men, Warriorism, and Mourning: The Development of Unconventional Warriors." *Pastoral Psychology* 66, no. 6 (December 1, 2017): 819–836.

Larson, Duane, and Jeff Zust. *Care for the Sorrowing Soul: Healing Moral Injuries from Military Service and Implications for the Rest of Us.* Eugene, OR: Cascade Books, 2017.

Lartey, Emmanuel Y. *In Living Color: An Intercultural Approach to Pastoral Care and Counseling.* 2nd ed. London: Jessica Kingsley Publishers, 2003.

Leslie, Kristen J. "Betrayal by Friendly Fire." In *War and Moral Injury: A Reader*, edited by Robert Emmet Meagher and Douglas A. Pryer, 245–255. Eugene, OR: Cascade Books, 2018.

Lifton, Robert Jay. *Home from the War: Learning from Vietnam Veterans.* New York: Other Press, 2005.

Litz, Brett T., and Jessica R. Carney. "Employing Loving-Kindness Meditation to Promote Self-and Other-Compassion among War Veterans with Posttraumatic Stress Disorder." *Spirituality in Clinical Practice* (July 12, 2018): 1–11.

Litz, Brett T., Leslie Lebowitz, Matt J. Gray, and William P. Nash. *Adaptive Disclosure: A New Treatment for Military Trauma, Loss, and Moral Injury.* New York: Guilford Press, 2016.

Litz, Brett T., Nathan Stein, Eileen Delaney, Leslie Lebowitz, William P. Nash, Caroline Silva, and Shira Maguen. "Moral Injury and Moral Repair in War Veterans: A Preliminary Model and Intervention Strategy." *Clinical Psychology Review* 29, no. 8 (2009): 695–706.

Machiavelli, Niccolò. *The Prince*, edited by Quentin Skinner and Russell Price. New York: Cambridge University Press, 1988.

Marx, Karl, and Friedrich Engels. *The Marx-Engels Reader*, edited by Robert C. Tucker, 2nd ed. New York: W.W. Norton & Company, 1978.

McDonald, Joseph, ed. *Exploring Moral Injury in Sacred Texts.* London: Jessica Kingsley Publishers, 2017.

Miller-McLemore, Bonnie J. *Christian Theology in Practice: Discovering a Discipline.* Grand Rapids: William B. Eerdmans Publishing Company, 2012.

———. "The Human Web: Reflections on the State of Pastoral Theology." *Christian Century* 110, no. 11 (April 7, 1993): 366–369.

Moon, Zachary. *Coming Home: Ministry That Matters with Veterans and Military Families*. St. Louis: Chalice Press, 2015.

———. "'Turn Now, My Vindication is at Stake:' Military Moral Injury and Communities of Faith." *Pastoral Psychology* 68, no. 1 (February 2019): 93–105.

———. *Warriors Between Worlds: Moral Injury and Identities in Crisis*. Lanham: Rowman & Littlefield Publishing Group, 2019.

Moore, S. K. *Military Chaplains as Agents of Peace: Religious Leader Engagement in Conflict and Post-Conflict Environments*. Lanham, MD: Lexington Books, 2014.

Morris, Joshua T. "'Thank You for Your Service:' Mapping Counter-Memories as a Form of Spiritual Care Support for Moral Injury." *Journal of Pastoral Theology* 28, no. 1 (2018): 34–44.

———. "Veteran Support and Antonio Gramsci: Counterhegemony as a Pastoral Theological Intervention." *Journal of Pastoral Theology*, 30, no. 3 (2020): 207–221.

Morton, Adam David. *Unraveling Gramsci: Hegemony and Passive Revolution in the Global Political Economy*. New York: Pluto Press, 2007.

Nash, William P., and Brett T. Litz. "Moral Injury: A Mechanism for War-Related Psychological Trauma in Military Family Members." *Clinical Child and Family Psychology Review* 16, no. 4 (2013): 365–375.

O'Brien, Tim. *The Things They Carried*. New York: Mariner Books, 2009.

Osmer, Richard R. *Practical Theology: An Introduction*. Grand Rapids, MI: William B. Eerdmans Publishing Company, 2008.

Pattison, Stephen. *Pastoral Care and Liberation Theology*. 2nd ed. London: SPCK Press, 1997.

Patton, John. *Pastoral Care in Context: An Introduction to Pastoral Care*. Louisville: Westminster John Knox Press, 1993.

Peters, David W. *Post-Traumatic God: How the Church Cares for People Who Have Been to Hell and Back*. New York: Morehouse Publishing, 2016.

———. "Sin Eater." In *War and Moral Injury: A Reader*, edited by Robert Emmet Meagher and Douglas A. Pryer, 208–218. Eugene, OR: Cascade Books, 2018.

Phillips, John A. *Christ for Us in the Theology of Dietrich Bonhoeffer*. New York: Harper & Row, 1967.

Pyne, Jeffrey M., Aline Rabalais, and Steve Sullivan. "Mental Health Clinician and Community Clergy Collaboration to Address Moral Injury in Veterans and the Role of the Veterans Affairs Chaplain." *Journal of Health Care Chaplaincy* (August 15, 2018): 1–19.

Ramshaw, Elaine. *Ritual and Pastoral Care*. Philadelphia: Fortress Press, 1987.

Rogers-Vaughn, Bruce. "Best Practices in Pastoral Counseling: Is Theology Necessary?" *Journal of Pastoral Theology* 23, no. 1 (2013): 1–26.

———. *Caring for Souls in a Neoliberal Age*. New York: Palgrave Macmillan, 2016.

———. "Class Power and Human Suffering: Resisting the Idolatry of the Market in Religious Practice and Pastoral Theology." In *Pastoral Theology and Pastoral Care: Critical Trajectories in Theory and Practice*, edited by Nancy J. Ramsay, 55–77. Malden, MA: Wiley Blackwell Publishers, 2018.

Root, Andrew. "Practical Theology: What Is It and How Does It Work." *Journal of Youth Ministry* 7, no. 2 (Spring 2009): 55–72.
Rushton, Cynda Hylton, Kathleen Turner, Rita Nakashima Brock, and Joanne M. Braxton. "Invisible Moral Wounds of the COVID-19 Pandemic: Are We Experiencing Moral Injury?" *AACN Advanced Critical Care* 32, no. 1 (2021): 119–125.
Saldaña, Johnny. *The Coding Manual for Qualitative Researchers*. 3rd ed. Los Angeles: Sage Publications, 2016.
Scheib, Karen D. *Pastoral Care: Telling the Stories of Our Lives*. Nashville: Abingdon Press, 2016.
Scott, A.O. "Review: 'American Sniper,' a Clint Eastwood Film With Bradley Cooper." *The New York Times*, December 24, 2014. Accessed April 17, 2017.
Segundo, Juan L. *Liberation of Theology*. Translated by John Drury. Eugene, OR: Wipf & Stock Publishers, 2002.
Shay, Jonathan. *Achilles in Vietnam: Combat Trauma and the Undoing of Character*. New York: Scribner, 1994.
———. "Moral Injury." *Intertexts* 16, no. 1 (2012): 57–66.
———. "Moral Injury." *Psychoanalytic Psychology* 31, no. 2 (2014): 182–191.
———. "Moral Leadership Prevents Moral Injury." In *War and Moral Injury: A Reader*, edited by Robert Emmet Meagher and Douglas A. Pryer, 301–306. Eugene, OR: Cascade Books, 2018.
Spivak, Gayatri Chakravorty. "Can the Subaltern Speak?" In *Colonial Discourse and Post-Colonial Theory: A Reader*, edited by Patrick Williams and Laura Chrisman, 66–111. New York: Columbia University Press, 1994.
Stake, Robert E. *The Art of Case Study Research*. Thousand Oaks: Sage Publications, 1995.
Suchocki, Marjorie Hewitt. *God, Christ, Church: A Practical Guide to Process Theology*. New York: Crossroad Publishing Company, 1995.
Sullivan, Winnifred Fallers. *A Ministry of Presence: Chaplaincy, Spiritual Care, and the Law*. Chicago: University of Chicago Press, 2014.
Tanielian, Terri, and Lisa H. Jaycox. *Invisible Wounds of War: Psychological and Cognitive Injuries, Their Consequences, and Services to Assist Recovery*. Santa Monica: RAND Corporation, 2008.
Thornton, Sharon G. *Broken yet Beloved: A Pastoral Theology of the Cross*. St. Louis: Chalice Press, 2002.
Tietz, Christiane. *Theologian of Resistance: The Life and Thought of Dietrich Bonhoeffer*. Translated by Victoria J. Barnett. Minneapolis: Fortress Press, 2016.
Trotsky, Leon. *1905*. New York: Vintage Books, 1971.
Townes, Emilie M. *Womanist Ethics and the Cultural Production of Evil*. New York: Palgrave Macmillan, 2006.
United Church of Christ. *Book of Worship*. Cleveland: United Church of Christ Press, 2012.
Vazquez-Torres, Jessica. "Does Moral Injury Have a Color? On Moral Injury and Race in the United States." Paper presented at the annual meeting of the American Academy of Religion. San Diego, CA. 2014.

Visco, Rosanne. "Postdeployment, Self-Reporting of Mental Health Problems, and Barriers to Care." *Perspectives in Psychiatric Care; Madison* 45, no. 4 (October 2009): 240–253.

Waggoner, Ed. "Taking Religion Seriously in the U.S. Military: The Chaplaincy as a National Security Asset." *Journal of the American Academy of Religion* 82, no. 3 (September 2014): 702–735.

Watkins Ali, Carroll A. *Survival and Liberation: Pastoral Theology in African American Context*. St. Louis: Chalice Press, 1999.

West, Cornel. "Black Theology and Marxist Thought." In *Black Theology: A Documentary History, Volume One: 1966–1979*, edited by James H. Cone and Gayraud S. Wilmore, 409–424. New York: Orbis Books, 1993.

White, Michael. *Maps of Narrative Practice*. New York: W.W. Norton & Company, 2007.

Whitlock, Craig, Leslie Shapiro, and Armand Emamdjomeh. "A Secret History of the War." *Washington Post*, December 9, 2019. Accessed July 21, 2020.

Wiinikka-Lydon, Joseph. "Mapping Moral Injury: Comparing Discourses of Moral Harm." *Journal of Medicine and Philosophy* 44, no. 2 (April 2019): 175–191.

———. *Moral Injury and the Promise of Virtue*. New York: Palgrave Macmillan, 2019.

———. "Moral Injury as Inherent Political Critique: The Prophetic Possibilities of a New Turn." *Political Theology* 18, no. 3 (2017): 219–232.

Index

absolution, in Liturgy of Solidarity, 135–38
Achilles in Vietnam (Shay), 4, 73
adaptive disclosure (AD), 31–32, 54
Adaptive Disclosure (Litz), 31–32
adaptive disclosure-enhanced (AD-E), 32–33, 54
"Afghanistan Papers," of *Washington Post*, 15
Ali, Carroll Watkins, 109–10
Althusser, Louis, 23, 50n66, 121; ideology critique of, 38–40, 93
American Sniper, 93–99, 120; ideology of, 96–99
Andrews, Dale P., 28–29, 49n21
anger, 30, 31
anxiety, 31
asymmetrical war, 27–28
authority, in MIEs, 4, 9, 28, 35, 73
Avanti! (Gramsci), 41

base, of Marx economic theory, 37
battle buddies: of Campbell, 84, 86, 108, 129; of Fisher, 1; of Lloyd, 69
believers: Campbell with, 85; in practical theology, 24
Bernstein, J. M., 28, 73–74
Bethge, Eberhard, 103, 139, 142
betrayal, in MIEs, 4, 9, 28, 35, 73
Black, indigenous, people of color (BIPOC), 2
Black Lives Matter, 3
Blacks: communal-contextual paradigm for, 109; liberation theology of, 55, 115n28, 140–41; solidarity of, 57
Black Theology and Black Power (Cone), 55
A Black Theology of Liberation (Cone), 55, 57
Boisen, Anton, 25; CPE and, 61–63

Bonhoeffer, Dietrich, 9, 18; communal-contextual paradigm and, 101–2; ecclesiological theory of, 101, 103; execution of, 139; on God, 85, 102, 105–6, 139–42; I-You of, 105, 106, 116n41; liberation theology and, 91, 101–2; sociality of, 102–7; theological anthropology of, 105–6
Brock, Rita Nakashima, 10, 29, 33–34, 145
BSS. *See* Building Spiritual Strengths
Buber, Martin, 116n41
Buddhism, 32, 86
Building Spiritual Strengths (BSS), 33, 54; confession and, 136
Buttigieg, Joseph, 123

Cabot, Richard, 61
Campbell, Phillip, 12; battle buddies of, 84, 86, 108, 129; communal-contextual paradigm for, 107–9, 110; dismissal of service by, 80; divided identity of, 79–83; driving by, 71–73; faith of, 84–86; God and, 84–85, 102, 105, 106, 140–41, 143; gratitude of, 79–80; IED and, 69, 70–71, 80–81, 82, 108; music and, 87–89, 133; neoliberalism of, 110–11, 113; prayer of, 86; PTSD of, 83, 86; reification of, 65–66; reintegration of, 75–83, 87–89, 128–29; religious community of, 87; SOPs and, 73, 80–81
caregivers, 4–5, 36; communal-contextual application and, 109–10; counterhegemony and, 128; liberative praxis for, 54, 59; living human web and, 7, 60, 62–63; MST and, 69; mutuality with care receiver, 62; neoliberalism and, 113
Caring for Souls in a Neoliberal Age (Rogers-Vaughn), 117n58
Carmichael, Stokely, 55
case study, 60; CPE on, 61–63
CBT. *See* cognitive behavioral therapy

Centre for Contemporary Cultural Studies (CCCS), 10, 38, 45–46
chaplains. *See* clergy
Chopp, Rebecca, 26
Chris Kyle (fictional character), 93–99
Christ existing as church-community (*Christus als Gemeinde existierend*), 103, 105
La Città Futura (Gramsci), 41
Clausen, Eric, 79
Clausewitz, Carl von, 121
clergy (chaplains): class role of, 124; as force multipliers, 17, 126–27; Gramsci on, 123–27; *locus theologiccus* of, 126
clinical pastoral education (CPE), 25; Boisen and, 61; on case study, 61–63
coercion: hegemony from, 43, 122; ISAs and RSAs and, 39; power of, 122
cognitive behavioral therapy (CBT): AD and, 32; for PTSD, 30
collaborative conversations, 34
color-blind theology, of white supremacy, 57
Combat and Operational Stress Control (Navy and Marine Corps), 29
Combat Support Hospital (CSH), 1, 67
Coming Home (Moon), 35
common sense: of Gramsci, 42, 51n81, 96; in hermeneutical circles, 57
communal-contextual paradigm: for Blacks, 109; Bonhoeffer and, 101–2; for Campbell, 107–9, 110; counterhegemony and, 127–30; feminist theology and, 109–10, 116n50; God in, 105–9; for MIEs, 18, 101–14; Segundo and, 102
community (*Gemeinschaft*), 106
Cone, James, 55, 57, 141
confession: in AD, 32; in Liturgy of Solidarity, 135–38
consent: encoding and decoding of, 96; Gramsci and, 96; hegemony from, 42–43

continuity of care, with VA, 13
correlation, in hermeneutical circles, 56
counterhegemony: caregivers and, 128; communal-contextual paradigm and, 127–30; encoding and decoding and, 96; of Gramsci, 9, 74, 119–23; Hall and, 44, 119–20; MIEs and, 119–45; oppositional readings and, 120–21; practical theology and, 121; reintegration and, 119; in Roman Catholic Church, 124–25; romanticizing of, 18
CPE. *See* clinical pastoral education
The Cross and the Lynching Tree (Cone), 141
CSH. *See* Combat Support Hospital

decoding, 95–96, 119–20
Democratic Party, 15
Denton-Borhaug, Kelly, 84
Department of Veterans Affairs (VA): adequate access to, 3; continuity of care with, 13
depression, 30, 31
Diagnostic and Statistical Manual of Mental Disorders (DSM), 9
Dingemans, Gijsbert, 24
disavowal, 42
Discipleship (Bonhoeffer), 103
divided identity, in reintegration, 79–83
divorce, of Gallagher, 76, 83
Doehring, Carrie, 35
Drescher, Kent D., 29
driving, by Campbell, Phillip, 71–73
DSM. *See Diagnostic and Statistical Manual of Mental Disorders*

Eagleton, Terry, 42
Eastwood, Clint, 96–97
ecclesiological theory, of Bonhoeffer, 101, 103
economic theory: of Marx, 37–38; of neoliberalism, 112; trickle down, 111–12
Ellsberg, Daniel, 15

empowering, in communal-contextual paradigm, 109, 110
encoding, 95–96; in *American Sniper*, 97
Engels, Friedrich, 37–38, 121
Enlightenment, 104; sociality and, 105
EOD. *See* Explosive Ordinance Disposal
Ethics (Bonhoeffer), 102, 103
The Exploration of the Inner World (Boisen), 61
Explosive Ordinance Disposal (EOD), 80–81, 108

faith: of Campbell, 84–86; in life verses, 84; Segundo on, 58
false consciousness, 38, 94
feminist theology, 8, 25, 33, 126; communal-contextual paradigm and, 109–10, 116n50; power and, 109
fiduciary assumption, of MIEs, 73
fight or flight, in PTSD, 30
Fisher, Lisa, 1–2, 12; battle buddies of, 1; God and, 85; meditative spaces of, 86; MST of, 69–70, 72–73; reification of, 66–67; reintegration of, 76, 77, 78, 85
FOBs. *See* forward operating bases
force multipliers, clergy as, 17, 126–27
forgiveness: in AD, 32; in BSS, 33; in Liturgy of Solidarity, 135–38
forward operating bases (FOBs), 70, 82
Foucault, Michel, 41
Frederick Zoller (fictional character), 98

Gallagher, Angela, 12; reification of, 67–68; reintegration of, 76, 83
Geertz, Clifford, 48n5
Gemeinschaft (community), 106
Gerkin, Charles V., 62
The German Ideology (Marx and Engels), 37–38
Gesellschaft (society), 106
Gingrich, Newt, 99
God: alienation from, 83–85; Bonhoeffer on, 85, 102, 105–7,

139–42; Campbell and, 84–85, 102, 105, 106, 140–41, 143; in communal-contextual paradigm, 107–9; Fisher and, 85; solidarity of, 143
God of the Oppressed (Cone), 141
Graeber, David, 51n81
Graham, Elaine, 25
Graham, Larry, 34–35, 36
Gramsci, Antonio, 23, 51n75; on clergy, 123–27; common sense of, 42, 51n81, 96; consent and, 96; counterhegemony of, 9, 74, 119–23; Hall on, 9, 11, 145n3; hegemony of, 42–43, 46; ideology critique of, 40–44, 93; intellectuals of, 43, 119, 121–27, 131; liberative praxis and, 41, 74, 87; Philip and, 146n14; war of position of, 18, 121–23
gratitude: of Campbell, 79–80; counterhegemony and, 129; of Lloyd, 77–78
Green, Clifford, 102
greetings, in Liturgy of Solidarity, 131–33
Greider, Kathleen, 116n50
Grelle, Bruce, 41
Il Grido del Popolo (Gramsci), 41
guiding, in communal-contextual paradigm, 109
guilt, 30, 48n10

Hall, Stuart, 3, 23, 52n97; counterhegemony of, 119–20; on encoding and decoding, 95–96, 97, 119–20; on Gramsci, 9, 11, 145n3; ideology critique of, 10–11, 44–46; ideology of, 18, 94–99; media and, 11, 46, 94–96; oppositional readings of, 94–99, 113, 120–21, 128; retotalization of, 128; on ruling class, 38; Segundo and, 58–59
Handlungswissenschaft (science of action), 24
Harnack, Adolf von, 104

Harris, Irene, 33, 136
head cases, 3, 27; Moon on, 35
healing, in communal-contextual paradigm, 109
hegemony, 5; from coercion, 43, 122; from consent, 42–43; encoding and decoding and, 96; of Gramsci, 42–43, 46; ideology and, 42–43, 58–59; intellectuals and, 122–23; of neoliberalism, 13, 110–13. *See also* counterhegemony
Helsel, Philip, 112
hermeneutical circles, of Segundo, 54, 55–59
heroes, 3, 27; Moon on, 35; snipers and, 98
Higher Power: in BSS, 33; confession and, 136
high-stakes situations, in MIEs, 4, 9, 27, 73
Hiltner, Seward, 109
historical materialism, 37
Hitler, Adolf, 104, 139
Hoggart, Richard, 45
Home from War (Lifton), 48n10
homelessness: of Fisher, 66–67; of Gallagher, 68, 76
Homer, 27, 73
Honneth, Axel, 28, 48n18

I and Thou (Buber), 116n41
ideological state apparatuses (ISAs), 39
ideological superstructure, in hermeneutical circles, 56
ideology: of *American Sniper*, 96–99; as contradictory, 3; defined, 10; of Hall, 18, 94–99; hegemony and, 42–43, 58–59; ISAs and RSAs and, 39; MIEs and, 93–99; as mythology of military, 3, 5; positive and negative connotations of, 42; of Segundo, 58; as system of ideas, 42; as unconscious, 94

"Ideology and Ideological State Apparatuses" (Althusser), 38–40, 50n66
ideology critique: of Althusser, 38–40, 93; of Gramsci, 40–44, 93; of Hall, 44–46; of Marx, 37–38, 93; MIEs and, 10–11, 36–46
IED. *See* improvised explosive device
The Iliad (Homer), 27, 73
improvised explosive device (IED), 69, 70–71, 80–81, 82, 108
inequality: neoliberalism and, 111–12; Rogers-Vaughn on, 112
Inglorious Basterds, 98–99
insomnia, 30
intellectuals: Bonhoeffer and, 103; of Gramsci, 43–44, 119, 121–27, 131; hegemony and, 122–23; power and, 125
interpellation, 39–40
ISAs. *See* ideological state apparatuses
I-You, of Bonhoeffer, 105, 106, 116n41

James, Henry, 44
Jinkerson, Jeremy, 30–31, 47
Johnson, Cedric, 111–12
Johnson, Gloria, 116n50

Kaepernick, Colin, 2
Keizer, Herman "Herm," Jr., 144–45
Kerry, John, 15
killed in action (KIA): family notification of, 14; Lloyd and, 69
Klay, Phil, 81
Kollektivperson, 106
Kutless, 87–89

Lacan, Jacques, 39
LaMothe, Ryan, 99, 112, 117n58, 144
Larson, Duane, 31
Lartey, Emmanuel, liberative praxis of, 17, 55, 59–60, 91
laser pointers, 71, 89n3
Latin America, liberation theology of, 55–56, 101, 115n28, 117n58

Lenin, Vladimir, 121
Leslie, Kristin, 116n50
Letters and Papers from Prison (Bonhoeffer), 101–2, 103, 139
Lettini, Gabriella, 10, 29, 33–34
liberating, in communal-contextual paradigm, 109, 110
Liberation of Theology (Segundo), 54–59
liberation theology, 5; of Blacks, 55, 115n28, 140–41; Bonhoeffer and, 91, 101–2; of Latin America, 55–56, 101, 115n28, 117n58; liberative praxis and, 8; MIEs and, 17–18, 54–55; practical theology and, 25–26
liberative praxis, 63, 143; for caregivers, 54, 59; Gramsci and, 41, 74, 87; of Lartey, 17, 55, 59–60, 91; liberation theology and, 8; practical theology and, 11–12, 53–60; religious community for, 87; ruling class and, 54; of Segundo, 54, 55–59; solidarity and, 8–9, 26
Life Together (Bonhoeffer), 104
life verses, 84
Lifton, Robert Jay, 48n10
Liturgy of Solidarity, 130–39, 141; absolution in, 135–38; confession in, 135–38; forgiveness in, 135–38; greetings in, 131–33; prayer in, 134–35; Prayer of Peace in, 138–39; story sharing in, 133–34
Litz, Brett, 3–4, 9–10, 28, 31–32
living human document, 61–63
living human web, of Miller-McLemore, 7, 60, 62–63
LKM. *See* loving-kindness meditation
Lloyd, Andrew, 12; antiwar perspective of, 74; battle buddies of, 69; PTSD of, 77, 83; reification of, 68–69; reintegration of, 77–78, 83, 128–29
locus theologiccus, 126
loving-kindness meditation (LKM), 32–33

Machiavelli, Niccolò, 43, 122
Maguen, Shira, 3–4, 29
malls, 81
Marx, Karl, 10, 23; economic theory of, 37–38, 46, 56; ideology critique of, 37–38, 93; liberative praxis of, 26; media and, 46. *See also* Althusser, Louis; Gramsci, Antonio
media: Hall and, 11, 46, 94–96; Marx and, 46
meditative spaces, 86; LKM, 32–33
Mercer, Joyce Ann, 112, 117n58
MIEs. *See* morally injurious events
military sexual trauma (MST), 69–70, 72–73; dehumanization from, 73–74
Miller-McLemore, 12, 24–25; living human web of, 7, 60, 62–63
Moon, Zachary, 3, 35, 47, 120
Moore, Michael, 98
Moore, Steve, 28
Moral Injury (Graham, L.), 34–35, 36
morally injurious events (MIEs), 2; audience of, 13; authority in, 4, 9, 28, 35, 73; betrayal in, 4, 9, 28, 35, 73; caregivers for, 36; communal-contextual paradigm for, 18, 101–14; community support for, 114; counterhegemony and, 119–45; events with, 9, 73; expanded definition of, 28; fiduciary assumption of, 73; future research on, 144–45; guilt with, 30; high-stakes situations in, 4, 9, 27, 73; historical context for, 26–35; ideology and, 93–99; ideology critique and, 10–11, 36–46; from lack of clear mission, 75; liberation theology and, 17–18, 54–55; literature on, 28; methodology and methods for, 11–13; naming of, 29; need for concept of, 29; organizational level of, 4; power and, 71–72; practical theology and, 23–36; racism and, 8, 29; as received or given, 34; reintegration and, 75–89; shame in, 30; similarities with PTSD, 30–31; subjective experience of, 4, 9; symptomology for, 30–31. *See also specific topics and individuals*
MST. *See* military sexual trauma
music, 87–89, 133
Mussolini, Benito, 40–41

Napoleon, 121
Nash, William, 3–4
national anthem, at sporting events, 2–3, 120
neoliberalism: of Campbell, 113; caregivers and, 113; hegemony of, 13, 110–13; inequality and, 111–12; oppositional readings and, 113; of Rogers-Vaughn, 113, 117n58
New Left, 38, 44–46
nightmares, 30
nurturing, in communal-contextual paradigm, 109

O'Brien, Tim, 133–34
Operators (special operators), 80
oppositional readings: counterhegemony and, 120–21; of Hall, 94–99, 113, 120–21, 128
L'ordine Nuovo (Gramsci), 41
organic intellectuals, 43, 123, 124
orthopraxis, 26, 55, 57
Osmer, Richard, 53–54; practical theology of, 48n2, 143

Pascal, Blaise, 40
patriotism, 2–9
Pattison, Stephen, 112, 117n58, 126
Patton, John, 108
PE. *See* prolonged exposure
"Pentagon Papers," 15
Peters, David W., 89n3
Philip, André, 146n14
Phillips, John, 103
Poling, James, 112
political parties, 15; clergy and, 123–24

The Popular Arts (Hall), 45
posttraumatic stress disorder (PTSD), 3, 4, 37, 54; of Campbell, 83, 86; CBT for, 30; in DSM, 9; fight or flight in, 30; of Lloyd, 77, 83; medication for, 83; Moon on, 120; PE for, 30; similarities with MIEs, 30–31
power: of coercion, 122; feminist theology and, 109; intellectuals and, 125; MIEs and, 71–72; in MST, 72–73
practical theology: believers in, 24; counterhegemony and, 121; democratization of, 25; liberation theology and, 25–26; liberative praxis and, 11–12, 53–60; MIEs and, 23–36; of Osmer, 48n2, 143; tasks of, 48n2, 53–54; *telos* and, 24
prayer, 86; Bonhoeffer on, 104; in Liturgy of Solidarity, 134–35
Prayer of Peace, in Liturgy of Solidarity, 138–39
Prison Notebooks (Gramsci), 41, 51n75, 123
prolonged exposure (PE), for PTSD, 30
PTSD. *See* posttraumatic stress disorder
Pyne, Jeffrey M., 49n37

Quaderni (Gramsci), 41, 123, 124
quasi-holy cause, 120

racism: MIEs and, 8, 29; Segundo on, 57
Ramsay, Nancy, 34, 112
rape, 28
Reagan, Ronald, 111–12
rear detachment, 76
reflexivity, 25
reification: of Campbell, 65–66; of Fisher, 66–67; of Gallagher, 67–68; of Lloyd, 68–69
reintegration: alienation of, 75–85; of Campbell, 75–83, 87–89, 128–29; counterhegemony and, 119; divided identity in, 79–83; of Fisher, 76, 77, 78, 85; of Gallagher, 76, 83; of Lloyd, 77–78, 83, 128–29; MIEs and, 75–89; music for, 87–89, 133; prayer and meditative spaces for, 86; religious community and, 86–87. *See also specific topics*
religious community, 86–87
re-membering, 108
repressive state apparatuses (RSAs), 39; in *Inglorious Basterds*, 98
Republican Party, 15
Reserve component, 12–13, 80; neoliberalism and, 112
retotalization, 128
Rogen, Seth, 98–99
Rogers-Vaughn, Bruce, 5; on inequality, 112; neoliberalism of, 113, 117n58
Roman Catholic Church: counterhegemony in, 124–25; Segundo and, 56
Root, Andrew, 25
RSAs. *See* repressive state apparatuses
ruling class: liberative praxis and, 54; Marx and, 38

Sanctorum Communio (Bonhoeffer), 9, 18, 101, 102–5, 116n41, 139
Sanders, Bernie, 51n81
Schucht, Tatiana, 41
science of action (*Handlungswissenschaft*), 24
Seeberg, Reinhold, 104
Segundo, Juan Luis, 5; communal-contextual paradigm and, 102; hermeneutical circles of, 54, 55–59; ideology of, 58; liberative praxis of, 54, 55–59
self-harm, 31
sexual harassment/assault. *See* military sexual trauma
Sexual Harassment Assault Response Prevention (SHARP), 67
shame, in MIEs, 30
SHARP. *See* Sexual Harassment Assault Response Prevention

Shay, Jonathan, 4, 9, 23, 26–27, 29, 31, 73
sites of resistance, 46
Smith, Archie, 112
sociality: of Bonhoeffer, 102–7; Enlightenment and, 105
society (*Gesellschaft*), 106
solidarity: of Blacks, 57; of God, 143; liberative praxis and, 8–9, 26. *See also* Liturgy of Solidarity
SOPs. *See* standard operating principles
Soul Repair (Brock and Lettini), 10, 29, 33–34
Soul Repair Center, 10, 34, 145
"Southern Question," 40
special operators (Operators), 80
sporting events, national anthem at, 120
standard operating principles (SOPs), 71, 73, 80–81
Stellvertretung (vicarious action), 106–7
story sharing, in Liturgy of Solidarity, 133–34
suicide, 120, 144
superstructure, of Marx economic theory, 37–38, 46, 56
Survival and Liberation (Ali), 109
suspicion, in hermeneutical circles, 56, 57
sustaining, in communal-contextual paradigm, 109
symmetrical war, 27–28

Tarantino, Quentin, 98–99
telos, 24
"thank you for your service," 120
theological anthropology, of Bonhoeffer, 105–6
thick description, 24, 48n5, 60, 65

The Things They Carried (O'Brien), 133–34
Thornton, Sharon, 8–9
torture, 28
traditional intellectuals, 43–44
trickle down economics, 111–12
Trotsky, Leon, 121, 145n7
Truth Commission on Conscience in War, 33

unconventional warriors, 99, 144
Unexploded Ordnance (UXO), 1
University of the Sacred Heart (University of Sacro Cuore), 125
UXO. *See* Unexploded Ordnance

VA. *See* Department of Veterans Affairs
Vazquez-Torres, Jessica, 8
vehicle-born IED (VBIED), 71–72
vicarious action (*Stellvertretung*), 106

Waggoner, Ed, 17, 127
Ward, Frances, 25
war of position, of Gramsci, 18, 121–23
Washington Post, "Afghanistan Papers" of, 15
Watson, Heather, 25
Weltanschauung, 41
West, Cornel, 115n28
white supremacy, 13; color-blind theology of, 57
Wiinikka-Lydon, Joseph, 4, 28–29, 73
Williams, Raymond, 45
womanist theology, 8, 25, 33, 109, 126

Žižek, Slavoj, 40
Zust, Jeff, 31

About the Author

Joshua T. Morris, PhD, is a bivocational scholar-practitioner. First, he teaches introductory and advanced spiritual care courses in hybrid and on-campus platforms at seminaries across the United States, and he also serves as a pediatric staff chaplain and also as a chaplain in the United States Army Reserve.

www.ingramcontent.com/pod-product-compliance
Lightning Source LLC
Chambersburg PA
CBHW020123010526
44115CB00008B/952